Multiple Imputation in Practice

With Examples Using IVEware

T0239574

Multiple Imputation in Practice

With Examples Using IVEware

Trivellore Raghunathan
Patricia A. Berglund
Peter W. Solenberger

CRC Press
Taylor & Francis Group
Boca Raton London New York

CRC Press is an imprint of the
Taylor & Francis Group, an **informa** business

A CHAPMAN & HALL BOOK

CRC Press
Taylor & Francis Group
6000 Broken Sound Parkway NW, Suite 300
Boca Raton, FL 33487-2742

First issued in paperback 2020

ISBN-13: 978-1-4987-7016-3 (hbk)
ISBN-13: 978-0-367-73403-9 (pbk)

Library of Congress Cataloging-in-Publication Data

Names: Raghunathan, Trivellore, author. | Berglund, Patricia A., author. |
Solenberger, Peter (Peter W.), author.
Title: Multiple imputation in practice : with examples using IVEware / by
Trivellore Raghunathan, Patricia Berglund, Peter Solenberger.
Description: Boca Raton, Florida : CRC Press, [2019] | Authors have developed
a software for analyzing mathematical data, IVEware. | Includes
bibliographical references and index.
Identifiers: LCCN 2018005310| ISBN 9781498770163 (hardback : alk. paper) |
ISBN 9781315154275 (e-book) | ISBN 9781498770170 (e-book (web pdf)) | ISBN
9781351650311 (e-book (epub) | ISBN 9781351640794 (e-book (mobi/kindle)
Subjects: LCSH: Missing observations (Statistics) | Multivariate analysis. |
Multivariate analysis--Data processing.
Classification: LCC QA276 .R2625 2019 | DDC 519.50285/53--dc23
LC record available at https://lccn.loc.gov/2018005310

Visit the Taylor & Francis Web site at
http://www.taylorandfrancis.com

and the CRC Press Web site at
http://www.crcpress.com

Contents

Preface

Almost every statistical analysis involves missing data, or rather missing elements in a data set (some prefer to use the term incomplete data). Such incomplete data may result from unit non-response where the selected subject or the survey unit refuses to provide any information or if the data collector was unable to contact the subject/unit. Incomplete data could also occur because of item non-response where the sampled subjects provide some information but not the others. The third type of situation can be called partial non-response as in, for example, drop-out in a longitudinal study at a particular wave. Another example of the partial response occurs in the National and Health and Nutrition Examination Survey (NHANES) which has three parts: a survey, medical examinations and laboratory measures. Some subjects may participate in a survey but refuse to participate in the medical examination (or not be selected for the medical examination) and/or refuse to provide a specimen for a laboratory analysis (or not be selected for the laboratory analysis). Finally, the incomplete data may occur due to design (planned missing data designs, split questionnaire designs etc), where not all subjects in the study/survey are asked every question.

There are three main approaches for dealing with missing data (or the analysis of incomplete data): (1) Weighting to compensate for nonrespondents (typically used for unit nonresponse but can also be used for handling some item nonresponse); (2) Imputation (typically used for item nonresponse but can be used for handling unit and partial nonresponse); and (3) Maximum likelihood or a Bayesian analysis based on the observed data under the posited model assumptions (such as multivariate normal, log-linear model etc).

Among these three approaches, imputation, specifically, multiple imputation, is the most versatile approach which can be implemented, relatively easily, as it capitalizes on widely available complete data analysis software. Under this approach, the missing set of values in a data set are replaced with several plausible sets of values. Each plausible set and observed set forms a completed data set (a plausible data set from the population). Each completed data set is analyzed separately. Relevant statistics (such as point and interval estimates, covariances, test statistics, p-values etc) are extracted from outputs of each analysis and are then combined to form a single inference.

Imputations are usually obtained as draws from a predictive distribution of the missing set of values, conditional on the observed set of values. There are number of ways to construct predictive distributions. One convenient approach is through a sequence of regression models, predicting each variable with missing values based on all other variables in the data set including auxiliary variables which may or may not be used in any particular analysis. This sequential regression or chained equation approach is flexible enough to handle different types of variables (continuous, count, categorical, ordinal, semi-continuous etc.) as well as complexities in a data set such as restrictions and bounds. Restrictions imply that some variables are relevant only for a subset of subjects (for example, the number of cigarettes smoked is only relevant for smokers; the years since quit smoking is relevant only for former smokers etc.). Bounds, on the other hand, imply logical or structural consistency between variables (for example, the number of years smoked for a smoker cannot exceed his/her age). Another example is a longitudinal study of children or adolescents, where the height of child in the subsequent wave cannot be less than the height in the previous wave.

The broad objective of this book is to provide a practical guide for multiple imputation analysis from simple to complex problems using real and simulated data sets. The data sets used in the illustration arise from a range of studies: Cross-sectional, retrospective, prospective and longitudinal studies, randomized clinical trials, and complex sample surveys. Simulated data sets are used to illustrate and explore certain methodological aspects.

Numerous tools have become available to perform multiple imputation as a part of popular statistical and programming packages such as R, SAS, SPSS and Stata. This book uses the latest version (Version 0.3) of *IVEware*, the software developed by the University of Michigan, that uses exactly the same syntax but works with all these packages as well as a stand-alone package (SR-CWARE) for performing multiple imputation analysis. Furthermore, *IVEware* can also perform survey analysis incorporating complex sample design features such as weighting, clustering and stratification.

Though *IVEware* is used to illustrate the applications of multiple imputation in a variety of contexts, the same analysis can be carried out using other packages that are built into Stata, R, SAS and SPSS. The book emphasizes analysis applications rather than software but using one common syntax system makes the presentation of the multiple imputation ideas easier. Not all features discussed in this book are built into *IVEware* but can be implemented using additional calculations (using another software or spreadsheet or electronic calculators) based on the output from *IVEware*. Sometimes a macro environment can be built to implement additional procedures. It is nearly impossible to develop a software that can calculate everything that one wants, especially, when the funding for developing software is rather scarce in an academic research setting.

This book consists of eleven chapters emphasizing the following nine areas of practical applications of MI:

1. Descriptive statistics: means, proportions and standard deviations; contrasts such as two sample comparisons of means and proportions;

2. Simple and multiple linear regression analysis;

3. Generalized linear regression models: logistic, Poisson, ordinal and multinomial logit models;

4. Time-to-event analysis, reliability and survival analysis;

5. Structural equation models;

6. Longitudinal data analysis for both continuous and binary outcomes;

7. Categorical data dnalysis, log-linear models;

8. Complex survey data analysis; and

9. Sensitivity analysis under a non-ignorable missing data mechanism.

A two semester sequence in statistics or biostatistics covering topics such as probability and distributions, statistical inference (both repeated sampling and Bayesian perspectives) and regression analysis including model diagnostics should provide sufficient background for most parts of the book. More detailed knowledge about the complete data model and subsequent analysis will be helpful (or even required) for other parts such as Structural Equation Models, or Survival Analysis, etc. even though the models are briefly described at the beginning of each chapter. A course in Bayesian analysis will also be helpful to understand the material.

This book is designed to be a companion to Raghunathan (2016) but also can serve as a companion to a general purpose book on analysis of incomplete data such as Little and Rubin (2002) and a book on multiple imputation Rubin (1987). Carpenter and Kenward (2013) is another useful companion book to thoroughly understand the fundamental concepts about multiple imputation. General concepts or theoretical results are only heuristically covered in this book. Another book that serves as a practical guide for performing multiple imputation using SAS is Berglund and Heeringa (2014) while guidance on using MICE in the R environment is Van Buuren (2012).

Additional readings are listed at the end of each chapter. They are not designed to be exhaustive references or a bibliography of missing data research. The listed references are not necessarily the original research articles but they might help further to understand the implementation and topics not covered in the listed books. Exercises at the end of each chapter are designed to extend the analysis presented or for additional practice with the techniques.

Both Raghunathan and Berglund have taught courses on missing data over the past several years. These range from one day to two semester long courses and workshops presented to a variety of audiences. They also also have consulted with many applied researchers on missing data issues. Solenberger is the chief architect of *IVEware* and its integration with SAS, R, Stata and SPSS. Together, we have tried to provide important insights into using multiple imputation for analyzing incomplete data. We are thankful to many students and collaborators for many questions, comments and challenges about one approach or the other, as these have shaped our course material and the presentation in the book. We are thankful to Dawn Reed who meticulously collected all the references and compiled them for this book. We hope that you find the book useful.

Trivellore Raghunathan ("Raghu")
Patricia Berglund
Peter Solenberger
Ann Arbor, Michigan

1

Basic Concepts

1.1 Introduction

Almost every statistical analysis involves missing data, or rather missing elements in a data set either due to unit non-response where the selected subject or the survey unit refuses to provide any information (also, if the data collector was unable to contact the subject) or due to item non-response where the sampled subjects provide some information but not the others. The third type of situation can be called partial non-response as in, for example, drop-out in a longitudinal study at a particular wave. Another example of the partial response occurs in the National and Health and Nutrition Examination Survey (NHANES) which has three parts: a survey, medical examinations and laboratory measures. Some subjects may participate in a survey but refuse to participate in the medical examination (or not be selected for the medical examination) and/or refuse to provide a specimen for a laboratory analysis (or not be selected for the laboratory analysis). Finally, the incomplete data may occur due to design (planned missing data designs, split questionnaire designs etc), where not all subjects in the study/survey are asked every question.

Before proceeding to discuss solutions for dealing with missing data, the the following three questions needs to be addressed:

1. What is a missing value?

2. Is there any pattern of the location of the missing values in data set?

3. Why are the values missing? What process lead to the missing values?

1.2 Definition of a Missing Value

A value for a variable in the analysis is heuristically defined as missing, if it is meaningful for a particular analysis and is hidden due to nonresponse. A clear cut example of such a variable is age of a sampled subject to be used as a covariate in a linear regression model. Now consider an example of a survey, where a question, "In the upcoming election, are you going to vote for the candidate B or F?", is asked leading to a variable X with response options (a) B, (b) F and (c) Don't know. One may develop an analysis plan that treats X as a three category analytical variable and, hence, there are no missing values in this analysis. On the other hand, suppose the analysis involves a projection of the winner in the election then subjects in the category (c) are to be treated as missing because their actual vote is a meaningful value for the analysis (projection of the winner) that is hidden.

Adding to this complexity, suppose that the following question is asked, "Are you planning to vote in the election?", leading to a variable Y with three response options (1) Yes, (2) No and (3) Don't know. Again, many analyses may use the variable Y as a three category analytical variable or a combination of X and Y as 9-response categories. For such analyses there are no missing values. However, for the projection of a winner or for the estimation of voter turnout rates, the "Don't knows" have to be treated as missing. Furthermore, the resolution of the missing values in Y determines the relevant population for X and then requires a resolution of missing values in X.

The above discussion illustrates that the missing value determination is specific to an analysis. That is, if the value of the variable is needed to construct an estimate/inference of an "Estimand" (the population quantity) then it is a missing value. One way of conceptualizing the missing value is to determine whether the value is hidden or should it be imputed for the analysis. In any case, even if the two variables X and Y are imputed for those in the "Don't know" categories, imputation flags should be created for reconstructing "meaningful" variables for any type of analysis and for diagnostic purposes.

1.3 Patterns of Missing Data

Consider a data matrix with n rows, representing subjects in the study, and p columns, representing variables measured on them. A pattern of missing data describes the location of the missing values in the data matrix. Sometimes, the rows and columns can be sorted to yield special patterns of missing data. Figure 1.1 illustrates three special patterns of missing data. Pattern (a) is the monotone pattern of missing data where Variable 2 is measured on a subset of people providing Variable 1, Variable 3 is measured on a subset of people providing Variable 2 etc. This type of pattern mostly occurs in a longitudinal setting where people who drop out are not followed in the subsequent waves. Pattern (b) describes a situation where a single variable is subject to missingness. This pattern may occur in a regression analysis with single covariate with missing values. Finally, Pattern (c) is common where two data sets are merged with some common variables and some variables unique to the individual data sets. This pattern can also be used to describe the causal inference data structure where the non-overlapping variables are the potential outcomes under treatments and the overlapping variables are pre-randomization or baseline variables. Pattern of missing data can be very useful to develop models or for

Figure 1.1: Illustrations of patterns of missing data

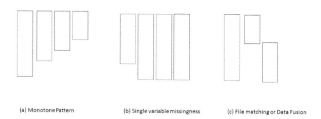

(a) Monotone Pattern   (b) Single variable missingness   (c) File matching or Data Fusion

modularizing the estimation task. For example, with the monotone pattern of missing data with four variables, the imputation model may be developed through the specifications of $Pr(Y_2|Y_1)$, $Pr(Y_3|Y_2, Y_1)$ and $Pr(Y_4|Y_1, Y_2, Y_3)$ which leads to imputing Y_2 just using Y_1, imputing Y_3 using observed or imputed Y_2 and Y_1, and, finally, imputing Y_4 using the observed or imputed Y_3, Y_2 and Y_1.

For Pattern (c) with three continuous variables Y_1, Y_2 and Y_3, there is no information about the partial correlation coefficient between Y_2 and Y_3 given Y_1. External information about this partial correlation coefficient will be needed to impute the missing values.

For most practical applications, the pattern of missing data will be arbitrary. The methods described in this book assume this general pattern of missing data, and, thus, can be used for these special patterns as well. However, imputation models may be simplified or augmented with external information when necessary, and will be indicated at appropriate places.

1.4 Missing Data Mechanisms

Once the missing values are identified for a specific analysis, the question arises "Why are they missing?" All methods for handling missing data make assumptions about the answer to this very key question. There is no assumption free approach for analyzing incomplete data. The assumptions about the reasons for missing values are conceptualized as the missing data mechanism.

Consider a simple example to understand the concept of the missing data mechanism. Suppose that Y_1 and Y_2 are two variables with Y_1 having no missing values and Y_2 having some missing values. The values of Y_2 are **M**issing **C**ompletely **A**t **R**andom (MCAR) if the probability of missingness does not depend on Y_1 or Y_2. In this case, the completers, that is, subjects with both Y_1 and Y_2 observed, are a random sub-sample of the full sample. Define $R_2 = 1$ for the respondents of Y_2 and $R_2 = 0$ for the nonrespondents. This assumption implies that the joint distribution of (Y_1, Y_2) satisfies the equality:

$$Pr(Y_1, Y_2|R_2 = 1) = Pr(Y_1, Y_2|R_2 = 0) = Pr(Y_1, Y_2).$$

An equivalent definition of MCAR mechanism is through the response propensity model, $Pr(R_2 = 1|Y_1, Y_2)$. Under MCAR, $Pr(R_2 = 1|Y_1, Y_2) = \phi$ where ϕ is an unknown constant. Thus, under MCAR, the complete case analysis provides unbiased information about the population, except for increasing in uncertainly due to the reduced sample size.

A weaker mechanism, called **M**issing **A**t **R**andom (MAR), assumes that the missingness in Y_2 may be related to Y_1 but not to the actual unobserved values of Y_2. One may state this assumption using the response propensity model as, $Pr(R_2 = 1|Y_1, Y_2, \phi) = Pr(R_2 = 1|Y_1, \phi)$ where ϕ are the unknown parameters. Under this assumption, $Pr(Y_2|Y_1, R_2 = 1) = Pr(Y_2|Y_1, R_2 = 0)$. Note that, this assumption is not verifiable based on the observed data because of lack of information to estimate $Pr(Y_2|Y_1, R_2 = 0)$.

An ignorable missing data mechanism places additional restrictions on the MAR or MCAR mechanisms. Suppose that $Pr(Y_1, Y_2|\theta)$ is the complete-data substantive model with the unknown parameter θ. Let Φ and Θ be the parameter spaces for the missing data mechanism parameter, ϕ, and the complete-data model parameter, θ, respectively. The two parameters θ and ϕ are called distinct, if there are no functional relationships between them. In other words, the joint parameter space of (θ, ϕ) is $\Theta \times \Phi$. Under the Bayesian framework, the distinctness implies that θ and ϕ are independent *a priori*.

The difference between MAR and ignorable missing data mechanisms is subtle. It is hard to conceive of a practical situation where the missing data are MAR or MCAR but the parameters are not distinct. Therefore, for all practical purposes, it will be assumed that when the data are MAR, the mechanism is also ignorable. If the mechanism is MAR (or MCAR) and the parameters are not distinct, then the implications of ignoring the mechanism on the inferences will have to be investigated on a case by case basis.

Finally, the missing data are **M**issing **N**ot **A**t **R**andom (MNAR) if the missingness depends upon the unknown responses. In this case, $Pr(R_2 = 1|Y_1, Y_2, \phi)$ is a function of Y_2 (perhaps, Y_1 as well). An implication of this assumption is that

$$Pr(Y_2|Y_1, R_2 = 1) \neq Pr(Y_2|Y_1, R_2 = 0).$$

Note that, this assumption, like MAR, is also not verifiable because Y_2 are not known whenever $R_2 = 0$. In this situation, an explicit model, $Pr(R_2 = 1|Y_1, Y_2, \phi)$, is needed, along with $Pr(Y_1, Y_2|\theta)$, for imputing the missing Y_2's.

In general, let Y be the complete data on p variables from n subjects arranged as a $n \times p$ matrix . Let R be the $n \times p$ response indicator matrix where a 1 indicates that the corresponding entry in Y is observed and 0, if it is unobserved. Let Y_{obs} be the collection of all the observed values in Y and Y_{mis} is the collection of all the missing values in Y. For a mechanism to be MCAR, one needs $Pr(R|Y, \phi) = \phi$, where ϕ is a constant; a MAR mechanism implies that $Pr(R|Y, \phi) = Pr(R|Y_{obs}, \phi)$ where ϕ is a vector or matrix of parameters; and for an ignorable missing data mechanism, ϕ and θ are distinct where θ is the parameter in, $Pr(Y|\theta)$, the data generating model. Since, $Pr(R|Y_{obs}, \phi)$ has no "informational" content about θ, conditional on knowing Y_{obs}, the mechanism can be ignored, or equivalently, can be completely unspecified.

The missing data mechanism is assumed to be ignorable for the most part, except in Chapter 10 where strategies for assessing sensitivity of inferences to this assumption are discussed. Generally, it is important to collect as many variables as possible that are correlated with variables with missing values to make the MAR assumption plausible and include them in the imputation process.

1.5 What is Imputation?

An imputation of an incomplete data set involves replacing the set of missing values by a plausible set of values (with due recognition while constructing the statistical inferences that these imputed values are not the actual values). The imputed data set, thus obtained is a plausible complete data set from the population. The imputed and observed sets of values should, therefore, under certain assumptions, inherently exhibit the properties of a typical complete sample data set from the population. To elaborate further, consider the same bivariate example discussed in the previous section. Assume that missing data in Y_2 are missing at random. One may be tempted to substitute all the missing values in Y_2, by their predicted value under a regression model. Suppose that the relationship is linear and the prediction equation is $\widehat{Y}_2 = \widehat{\beta}_o + \widehat{\beta}_1 Y_1$ where $\widehat{\beta}_o$ and $\widehat{\beta}_1$ are the ordinary least square estimates of the intercept and slope, respectively. In this imputed data set, all the observed values of Y_2 will typically be spread around the regression line but all of the imputed values will be exactly on the regression line, thus not a plausible complete data set from the population.

To heuristically understand the concept of a plausible complete data set, consider a scatter plot of Y_2 against Y_1 in the imputed data set with observed and imputed set of values distinguished through different colors or symbols. The perspective in this book is that for the imputed data set to be a plausible completed data set, the two colors or symbols should be "exchangeable", for any Y_1 value. That is, the labeling of observed or imputed should be at random, at any given value of Y_1.

The most principled approach for creating imputations is through using Bayesian principles. Suppose that the bivariate data are arranged such that complete-cases are the first r subjects and incomplete cases are the last $n - r$ subjects where n is the sample size. Assume a normal linear regression model for Y_2 on Y_1, $Y_2 = \beta_o + \beta_1 Y_1 + \epsilon$, $\epsilon \sim N(0, \sigma^2)$. Writing X_c as a $r \times 2$ matrix of predictors (a column of ones and Y_1 values) for the $r \times 1$ outcome variable, Y_{2c}, based on the r complete cases, $\beta = (\beta_o, \beta_1)^T$ as the column vector of the regression coefficients (the super script T denotes the matrix transpose) and a Jeffreys prior $Pr(\beta, \sigma) \propto \sigma^{-1}$, the following results are from the standard Bayesian analysis:

1. The posterior distribution of σ^2 is given by $(r - 2)\widehat{\sigma}_c^2 / \sigma^2 | \widehat{\sigma}_c^2 \sim \chi_{r-2}^2$.

2. The posterior distribution of β is given by

$$\beta | \widehat{\beta}_c, \sigma, X_c \sim N(\widehat{\beta}_c, \sigma^2 (X_c^T X_c)^{-1}),$$

where $\widehat{\beta}_c = (X_c^T X_c)^{-1} X_c^T Y_{2c}$ is the ordinary least squares complete-case estimate of β and $\widehat{\sigma}_c^2 = (Y_{2c} - X_c \widehat{\beta}_c)^T (Y_{2c} - X_c \widehat{\beta}_v) / (r - 2)$ is the corresponding residual variance.

The following steps describe the generation of imputations as draws from the posterior predictive distribution of the missing set of values conditional on the observed set of values. First, generate a chi-square random variable, u, with $r - 2$ degrees of freedom and define $\sigma_*^2 = (r - 2)\widehat{\sigma}_c^2 / u$. Next, generate β^* from a bivariate normal distribution with mean $\widehat{\beta}_c$ and the covariance matrix $\sigma_*^2 (X_c^T X_c)^{-1}$. Finally, generate imputations, Y_{2i}^*, for each missing value Y_{2i} from a normal distribution $N(\beta_o^* + \beta_1^* Y_{1i}, \sigma_*^2)$ where $i = r + 1, r + 2, \ldots, n$.

The above procedure incorporates uncertainty in the value of the parameters (β, σ) as well as the uncertainty in the value of the response variable, Y_2 for the nonrespondents. This is the proper approach for creating imputations from the Bayesian perspective. When the fraction of missing values is small, the uncertainty in the parameter may be relatively small and the imputation can be simplified by drawing values from a normal distribution $N(\widehat{\beta}_{oc} + \widehat{\beta}_{1c} Y_{1i}, \widehat{\sigma}_c^2)$, though it is not proper.

The regression model need not be linear but can be developed based on substantive knowledge and empirical investigations. The distribution of the residuals need not be normal. One strategy for a non-normal outcome is to transform Y_2 to achieve normality, impute values on the transformed scale and then re-transform to the original scale. One has to be careful in applying this strategy because an imputed value may appear reasonable on the transformed scale but, when re-transformed, may be unreasonable on the original scale. For example, suppose that Y_2 is income and a logarithmic transformation is used for the modeling purposes. An imputed value from the tail of the distribution on the logarithmic scale when exponentiated may lead to an unreasonable imputation.

A better option, perhaps, is to use a non-normal distribution for the residuals in a model on the original scale. An example of such a distribution is Tukey's gh-distribution that can accommodate a variety of departures from the normal distribution.

A random variable, X, follows Tukey's gh distribution with location μ, scale σ, skewness parameter $g(0 \leq g \leq 1)$ and the kurtosis or elongation parameter $h(0 \leq h \leq 1/2)$ if

$$X = \mu + \sigma \times Z \times \frac{\exp(gZ) - 1}{gZ} \times \exp(hZ^2/2)$$

where Z is a standard normal random variable. The parameter g controls skewness and h controls the thickness of the tail of the distribution. For modeling of the residuals, $\mu = 0$. The three unknown parameters cover a wide variety of non-normal distributions.

For given values of the parameters, (μ, σ, g, h), it is fairly easy to draw imputations (since X is a transformation of a standard normal deviate). However, developing the likelihood function and computing the maximum likelihood estimates of the parameters, (μ, σ, g, h), are computationally complex. The quantile based estimates are relatively easy to compute, and may even be more attractive given their robustness properties. Specifically, let Q_p denote the quantile corresponding to probability p. That is, $Pr(X \leq Q_p) = p$. Let Z_p be the corresponding standard normal distribution quantile. An obvious estimate of μ is the median $Q_{0.5}$.

It is easy to show that,

$$A_p = \frac{Q_{1-p} - Q_{0.5}}{Q_{0.5} - Q_p} = exp(-gZ_p).$$

Suppose $\widehat{Q}_p$ are the estimates for a selected set of values of p. The negative of the slope of the regression of $\log \widehat{A}_p$ on Z_p through the origin can be used as an estimate of g.

Similarly, it can be shown that

$$B_p = \frac{g(Q_{1-p} - Q_{0.5})}{\exp(-gZ_p) - 1} = \sigma \exp(hZ_p^2/2).$$

Thus, regressing $\log \widehat{B}_p$ on $Z_p^2/2$ results in the intercept $\log \widehat{\sigma}$ and slope $\widehat{h}$.

For imputation purposes, use the ordinary least squares (or any other robust regression technique) approach to estimate β_o and β_1, construct the residuals and their quantiles and then fit the two regression models described above to obtain, $\widehat{g}, \widehat{h}$ and $\widehat{\sigma}$ (estimate of μ for the residuals is 0). The imputations are defined as:

$$Y_{2i}^* = \widehat{\beta}_o + \widehat{\beta}_1 Y_{1i} + \widehat{\sigma}\widehat{g}^{-1}(\exp(\widehat{g}Z_i) - 1)\exp(\widehat{h}Z_i^2/2) \tag{1.1}$$

where $Z_i, i = r + 1, r + 2, \ldots, n$ are independent standard normal deviates. Note that this approach is not proper, because uncertainty in the estimates of the parameters $(\beta_o, \beta_1, \sigma, g, h)$ is not incorporated in the imputation process. A simple fix, is to use a bootstrap sample of the original data, estimate the parameters as outlined above using the bootstrap data and then apply the imputation process described in Equation (1.1).

Imputations can also be created based on data driven techniques such as a semiparametric model, using, for example, splines (not discussed in this book) or a totally nonparametric approach as described below:

1. Create H strata based on all the n values Y_1 and suppose that r_h and m_h are the number of respondents and nonrespondents in stratum $h = 1, 2, \ldots, h$. Let $Y_{2jh}, j = 1, 2, \ldots, r_h$ denote the observed responses in stratum h.

2. Apply a Bayesian bootstrap within each stratum to sample m_h missing values from the r_h observed values as imputations. The following are the steps for the Bayesian bootstrap:

   (a) Draw, $r_h - 1$ uniform random numbers between 0 and 1 and order them, $u_o = 0 \leq u_1 \leq u_2 \ldots \leq u_{r_h-1} \leq 1 = u_{r_h}$.

   (b) Draw another uniform random number, v and select Y_{2jh} as the imputation for a missing value if $u_{j-1} \leq v \leq u_j$ where $j = 1, 2, \ldots, r_h$.

   (c) Repeat the step (b) above for all the m_h missing values.

   (d) Repeat the steps from (a) to (c) for all strata.

Instead of the Bayesian Bootstrap (Steps (a) -(c)), an approximate version given below may be easy to implement.

(a*) Sample r_h values with replacement from $Y_{2jh}, j = 1, 2, \ldots, r_h$ and denote the resampled values as $Y_{2jh}^*, j = 1, 2, \ldots, r_h$.

(b*) Sample m_h values with replacement from the resampled values in (a*).

The underlying theory behind the Bayesian Bootstrap is that the distinct values of the observations are modeled using a multinomial distribution with unknown cell probabilities and a non-informative Dirchelet prior distribution for these cell probabilities.

There are numerous other possibilities. For example, Tukey's gh distribution could be used within each stratum defined by Y_1 instead of the Approximate Bayesian bootstrap. The key is to construct a sensible, good fitting prediction model for Y_2 given Y_1 and then use the model to generate imputations.

1.6   General Framework for Imputation

Taking the cue from the specific bivariate example, consider now a more general framework for creating model-based imputations. Suppose that the complete data can be arranged as a $n \times p$ matrix, Y, with rows corresponding to n subjects and p columns corresponding to variables. Let R be $n \times p$ matrix with 1's and 0's where 1 indicates that the corresponding elements in Y are observed and 0 indicates missing. A statistical model is the joint distribution of (Y, R) which can be partitioned as $Pr(Y|\theta) \times Pr(R|Y, \phi)$ or as $Pr(Y|R, \alpha)Pr(R|\beta)$ where (θ, ϕ) and (α, β) are the unknown parameters. The former is called the "selection" model formulation, and the latter is the "pattern-mixture" model.

Let Y_{obs} denote all the elements in Y for which the corresponding elements in R are equal to 1. Similarly, let Y_{mis} be the collection of all elements in Y corresponding to entries in R equal to 0. With a slight abuse of notation, write $Y = (Y_{obs}, Y_{mis})$.

Consider the selection model formulation and suppose that $\pi(\theta, \phi)$ is the prior density for the unknown parameters (θ, ϕ) and let $f(y|\theta)$ and $g(r|y, \phi)$ be the corresponding probability densities (or mass functions) for Y and R, conditional on Y. The goal of the imputation is to draw from the predictive distribution $Pr(Y_{mis}|Y_{obs}, R)$ with the density

$$f(y_{mis}|y_{obs}, r) = \frac{\int f(y_{obs}, y_{mis}|\theta)g(r|y_{obs}, y_{mis}, \phi)\pi(\theta, \phi)d\theta d\phi}{\int \int f(y_{obs}, y_{mis}|\theta)g(r|y_{obs}, y_{mis}, \phi)\pi(\theta, \phi)d\theta d\phi dy_{mis}},$$

where y_{obs} and r are the observed values of Y_{obs} and R in the data set being analyzed.

Note that, under the missing at random mechanism, $g(r|y_{obs}, y_{mis}, \phi) = g(r|y_{obs}, \phi)$. The distinctness condition implies $\pi(\theta, \phi) = \pi_1(\theta)\pi_2(\phi)$. Thus, for the ignorable missing data mechanism,

$$f(y_{mis}|y_{obs}, r) = \frac{\int f(y_{obs}, y_{mis}|\theta)\pi_1(\theta)d\theta}{\int \int f(y_{obs}, y_{mis}|\theta)\pi_1(\theta)d\theta dy_{mis}}.$$

That is, the model for the response indicator R (the missing data mechanism) can be ignored (or unspecified) while constructing the predictive distribution of the missing set of values for the imputation purposes. Through out the book (except in Chapter 10) missing data mechanisms are assumed to be ignorable. However, the imputation based approach can be used for nonignorable missing data mechanisms.

1.7 Sequential Regression Multivariate Imputation (SRMI)

The general framework described in the previous section, even under the ignorable missing data mechanism, is difficult to implement when the data set has a large number of variables of different types (continuous, count, categorical, ordinal, censored etc.), structural dependencies (such as years smoked cannot exceed the age of the subject or lab values have to be between certain logical bounds etc.) and restrictions (some variables makes sense only for a section of a sample such as years smoked is not applicable for never smokers). Constructing a joint distribution for all the variables with missing values with all such complexities is nearly impossible.

The sequential regression (or chained equations, flexible conditional specifications) approach uses a sequence of regression models where each variable with missing values is regressed on all other variables as predictors. A Gibbs sampling style iterative approach is used to draw values from the posterior predictive distribution of the missing values under each regression model. Suppose that U is a $n \times q$ matrix of q variables on n subjects with no missing values and $Y_1, Y_2, \ldots, Y_p$ are the p variables with some missing values. It is not necessary to have any variables with no missing values. That is, U could be a column of 1s representing the intercept term.

For the first iteration, Y_1 is regressed on U to impute the missing values in Y_1 resulting in a completed $n \times 1$ vector, $Y_1^{(1)}$, of observed and imputed values. A fully Bayesian approach is used (similar to bivariate example discussed previously). Next, Y_2 is regressed on $U, Y_1^{(1)}$ to obtain $Y_2^{(1)}$. The algorithm continues until the missing values in Y_p are imputed by regressing it on $U, Y_1^{(1)}, Y_2^{(1)}, \ldots, Y_{p-1}^{(1)}$.

If the missing data pattern in the sequence of p variables is monotone then one can stop with this first iteration. If the pattern is not monotone then the imputed values in Y_1, for example, does not depend on the observed values in the subsequent variables. Thus, the subsequent iterations use all the variables as predictors. For example, the missing values in Y_1 are re-imputed by regressing it on $U, Y_2^{(1)}, Y_3^{(1)}, \ldots, Y_p^{(1)}$ to yield the next generation, $n \times 1$ vector, $Y_1^{(2)}$, of the observed and imputed values. Similarly, the missing values in Y_2 are re-imputed by regressing it on $U, Y_1^{(2)}, Y_3^{(1)}, \ldots, Y_p^{(1)}$. Thus, at iteration t, the missing values in Y_j are re-imputed by regressing the observed values of Y_j on $U, Y_1^{(t)}, \ldots, Y_{j-1}^{(t)}, Y_{j+1}^{(t-1)}, \ldots, Y_p^{(t-1)}$.

The imputation procedure is now described in a general form but later adapted to a specific regression model. Let $v_i, i = 1, 2, \ldots, r$ be observed values of a variable to be imputed (one of the Y's above). Let z_i be the corresponding vector of predictors and $z_j, j = r+1, r+2, \ldots, n$ be the vector of predictors corresponding to missing values in v. Note that z's will consist of observed and imputed values of all other variables. Let $E(v_i | z_i) = \mu_i =$

$h(z_i^T \beta)$ and $Var(v_i | z_i) = \sigma^2 g(\mu_i)$ be the mean and variance of the conditional distribution, $Pr(v | z, \beta, \sigma)$, a member of the exponential family. Assume a flat prior, $\pi(\beta, \sigma) \propto \sigma^{-1}$. Let $\pi(\beta, \sigma | \{v_i, z_i, i = 1, 2, \ldots, r\})$ be the posterior density of the unknown parameters. The following procedure is used to impute the missing values:

1. Draw a value of (β, σ), say, (β^*, σ^*) from its posterior distribution.

2. Draw missing v_j from its conditional distribution $Pr(v | z_j, \beta^*, \sigma^*)$.

The regression model depends upon the variable being imputed such as normal linear (in the original or transformed scale) for continuous variables, logistic or probit for binary, Poisson for count, multinomial logit for categorical etc. The key is to obtain well fitting regression models for each variable. The imputation problem reduces to developing p regression models as building blocks.

Many data sets contain a mixed type of variable that is essentially continuous but with a spike at 0. An example of such a variable is real estate income. The value is 0 for many subjects in the study (those without income generating real estate) and a positive or negative value for the rest. These types of variables can be handled by using a two stage imputation process. The first stage imputes zero/non-zero status using a logistic regression model and then, conditional imputation using a continuous variable approach to impute non-zero values.

A normal distribution may not be a reasonable model for a continuous variable, even on the transformed scale. The following non-parametric approach (extending the Approximate Bayesian Bootstrap approach for the bivariate data example) may be used for such variables. Suppose that Y is a continuous variable with missing values and let R be the corresponding response indicator or dummy variable. Let Z be the predictors. The imputation approach has the following two steps:

1. Balance the respondents and non-respondents on Z. Run a logistic regression of R on Z and obtain the propensity scores, $\widehat{P}$. Run a regression of Y on Z to obtain predicted values $\widehat{Y}$. Create H classes or strata through a stratification based on the two variables $\widehat{P}$ and $\widehat{Y}$.

2. Imputation is performed within each stratum. Let r_h and m_h be the number of observed and missing values in stratum $h = 1, 2, \ldots, H$. Either apply an approximate Bayesian Bootstrap or Tukey's gh approaches (described in Section 1.4) to impute the missing values.

The number of classes or strata depends upon the sample size, the number of respondents and non-respondents. Typically, five classes based on the propensity scores, $\widehat{P}$, removes most of the bias (90%) due to imbalances in the covariates, Z, between the respondents and non-respondents. Further stratification of each propensity score class into five subclasses based on $\widehat{Y}$ may create homogeneous group of respondents and non-respondents. Thus the quintiles of the propensity scores and quintiles of the prediction score within each quintile class are used as a default in *IVEware* (software used to illustrate all the examples discussed in this book). This strategy is mainly derived for large data sets. A smaller number of classes may be chosen depending upon the sample size.

The restrictions are handled by fitting the data to a relevant subset. Subjects who do not satisfy the restriction are treated as a separate category when the restricted variables are subsequently used as predictors. The bounds, applicable only to continuous or semi-continuous variables, are handled by drawing values from a truncated distribution where the truncation is determined by the lower and upper bound specified by the user. These bounds can be a scalar of variables in the data set. These features are illustrated through concrete examples in the later chapters.

When the number of variables is large, the regression model building can be time consuming. Some dimensionality reduction features are built into *IVEware*. One is through specifying a minimum additional R^2 that is needed for a variable to be included in the model in a stepwise variable selection procedure. Another is through specifying the maximum number of predictors to be included in the model. Both features can be specific to a variable or global specifications.

When building the prediction models, interactions and non-linear terms may have to be included. As these are derived variables, the individual variables are imputed first and then derived variables are constructed to be included as predictors. The interactions can be specific to a variable being imputed or a global declaration where the interactions are included in all the models (except in the models for predicting the variables in the interaction/nonlinear term). Additional features in *IVEware* will be illustrated through examples.

The key step is to obtain well fitting regression models for each variable with missing values. The model building task is an iterative process where exploratory data analysis and substantive theory might suggest a working model. Model diagnostics such as residual plots, histograms of the residuals and other procedures are used to refine the model. The refined model is assessed for how well it fits and further refinements are made.

1.8   How Many Iterations?

Imputations using SRMI are created through a Gibbs style iterative process and the question arises of how many iterations should be used to achieve stable imputations. Many empirical studies show that after 5 to 10 iterations, no further mixing is needed to garner the predictive power of the observed data to create imputations. One potential approach is to monitor convergence of key statistics with the number of iterations. One general statistic is the empirical joint moment generating function of the variables in the data set. Let $Y^l = (Y_1^l, Y_2^l, \ldots, Y_p^l)$ be the lth completed data set on p variables where Y_i^l either has the observed or imputed values. Let $t = (t_1, t_2, \ldots, t_p)$ be any arbitrary vector of numbers centered around 0 (say, uniform numbers between -1 and 1). Define

$$M_t(Y^l) = \sum_j \left(\sum_i \exp\left(t_j Y_{ij}^l\right) / n \right)$$

where the sum is taken over both, p variables and n subjects in the completed data set. The multiple imputation estimate of the moment generating function is $M_t = \sum_l M_t(Y^l)/M$. The number of iterations can be determined adaptively, by constructing this scalar quantity for several randomly chosen values of t to determine the number of iterations.

1.9 A Technical Issue

Note that, the SRMI approach requires only specifications of the conditional distributions of the variables with missing values and allows enormous flexibility in developing these basic regression models. A technical issue arises because the specifications of these conditional distributions may not correspond to a joint distribution. Consider an example with two variables, X and Y, with missing values and the following two models seem to fit the data well,

$$f(x|y) \sim Normal(\alpha_o + \alpha_1 y + \alpha_2 y^2, \sigma_1^2),$$

and

$$g(y|x) \sim Normal(\beta_o + \beta_1 x + \beta_2 x^2 + \beta_3 x^3, \sigma_2^2).$$

These two models are incompatible as no joint distribution, $h(x,y)$, exists with, f and g, as conditional distributions.

Though the SRMI algorithm resembles the standard Gibbs sampler, the convergence properties are not well understood when the underlying conditional distributions are not compatible with a joint distribution. Liu et al.(2014), Hughes et al. (2014) and Zhu and Raghunathan (2015) provide various regularity conditions for the algorithm to converge. Various simulation studies in Zhu and Raghunathan (2015) and Zhu (2016) show that the SRMI procedure produces valid inferences (in terms of bias, mean square error and confidence coverage properties) when well fitting models are used for each conditional distribution. Thus, the practical implication of this technical issue is not clear, and may not be important, if well fitting models are used in the imputation process.

1.10   Three-variable Example

1.10.1   SRMI Approach

To illustrate the model development process, consider a data set with three variables, (Y_1, Y_2, Y_3). Suppose that all three variables have missing values in some arbitrary pattern. For creating imputations, three regression models are needed: (1) Y_1 on (Y_2, Y_3); (2) Y_2 on (Y_1, Y_3) and (3) Y_3 on (Y_1, Y_2). The three scatter plots in Figure 1.2 are the first step in the model building task.

    The two scatter plots in Figure 1.2 indicate nonlinear or non-additivity relationships between Y_1 and Y_3 and Y_2 and Y_3, respectively, and the other suggests a linear relationship between Y_1 and Y_2. The model building tasks begin with fitting the following three models:

1. $y_1 = \alpha_o + \alpha_1 y_2 + \alpha_3 y_3 + \alpha_3 y_2 y_3 + \epsilon_1$ where $\epsilon_1 \sim N(0, \sigma_1^2)$;

2. $y_2 = \beta_o + \beta_1 y_1 + \beta_2 y_3 + \beta_3 y_1 y_3 + \epsilon_2$ where $\epsilon_2 \sim N(0, \sigma_2^2)$; and

3. $y_3 = \gamma_o + \gamma_1 y_1 + \gamma_2 y_2 + \gamma_3 y_1 y_2 + \epsilon_3$ where $\epsilon_3 \sim N(0, \sigma_3^2)$.

The residual diagnostics suggest that Models 1 and 2 can be improved by adding the square term y_2^2 and y_1^2, respectively. The resulting final imputation models are $y_1 = \alpha_o + \alpha_1 y_2 + \alpha_2 y_3 + \alpha_3 y_2 y_3 + \alpha_4 y_2^2 + \epsilon_1$, $y_2 = \beta_0 + \beta_1 y_1 + \beta_2 y_3 + \beta_3 y_1 y_3 + \beta_4 y_1^2 + \epsilon_2$ and $y_3 = \gamma_o + \gamma_1 y_1 + \gamma_2 y_2 + \gamma_3 y_1 y_2 + \epsilon_3$.

1.10.2   Joint Model Approach

A more principled approach for creating multiple imputations is to develop a joint distribution for (Y_1, Y_2, Y_3) and then construct the needed predictive distributions. Given the complexity of the relationship among these variables, developing such a joint distribution is not an easy task. One might attempt to build such a model as follows:

1. A histogram of y_1 suggests a model $y_1 | \mu, \sigma^2 \sim N(\mu_1, \sigma_1^2)$.

2. The scatter plot in Figure 1.2 suggests that $y_2 | y_1, \theta, \sigma_2 \sim N(\theta_o + \theta_1 y_1, \sigma_2^2)$ where $\theta = (\theta_o, \theta_1)$.

3. The regression analysis suggests that $y_3 | y_1, y_2, \phi, \sigma_3 \sim N(\phi_o + \phi_1 y_1 + \phi_2 y_2 + \phi_3 y_1 y_2, \sigma_3^2)$ where $\phi = (\phi_o, \phi_1, \phi_2, \phi_3)$.

Thus, the proposed joint distribution is

$$f(y_1, y_2, y_3 | \omega) = (2\pi)^{-3/2} \sigma_1^{-1} \sigma_2^{-1} \sigma_3^{-1}$$

$$\exp\left[-\left\{ \left(\frac{y_1 - \mu_1}{\sigma_1} \right)^2 + \left(\frac{y_2 - \theta_o - \theta y_1}{\sigma_2} \right)^2 + \left(\frac{y_3 - \phi_o - \phi_1 y_1 - \phi_2 y_2 - \phi_3 y_1 y_2}{\sigma_3} \right)^2 \right\} \right]$$

$$(1.2)$$

where $\omega = (\mu_1, \theta, \phi, \sigma_1, \sigma_2, \sigma_3)$.

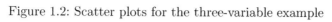

Figure 1.2: Scatter plots for the three-variable example

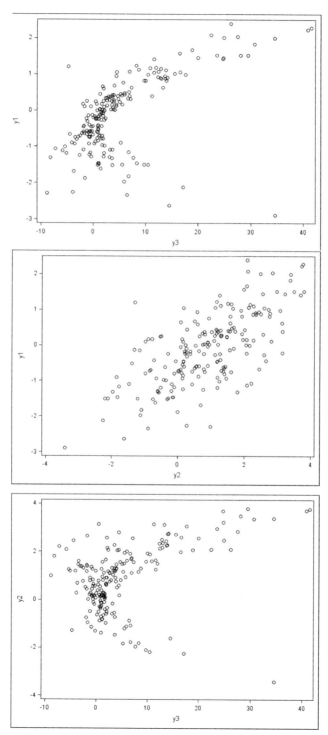

Assume a flat prior for w, $Pr(w) \propto (\sigma_1 \sigma_2 \sigma_3)^{-1}$. A Gibbs sampling algorithm could be used to generate imputations as follows:

1. Generate the initial missing values using some random mechanism

2. Generate the value of w from its posterior distribution given the complete data. This is fairly easy to do as three distinct analyses: marginal model of y_1 (for generating (μ_1, σ_1)), the regression of y_2 on y_1 (for generating (θ, σ_2)) and the regression of y_3 on y_1 and y_2 (for generating (ϕ, σ_3))

3. Regenerate the missing values from the appropriate predictive distribution given the parameters generated in Step 2 (See Step 2a below).

4. Iterate between Steps 1, 2 and 3 until the convergence is attained.

An alternative approach for starting the Gibbs cycle is to generate initial values of the parameters as given below:

1a. Generate the initial value of the parameter w, by drawing a bootstrap sample and then estimate w based on the complete cases in the bootstrap sample.

2a. Generate the imputations from the predictive distribution given the observed values and the initial value of the parameters.

3a. Generate the new value of the parameters as in Step 2 above.

4a. Iterate between Steps 1a, 2a and 3a until the convergence is attained.

For Step 2 (or Step 3a), the following itemized list provides details about the drawing the value of the parameters given the complete data:

- Let $\bar{y}_1$ and s_1 be the mean and standard deviations of variable Y_1 in the completed data set obtained in Step 1. Generate σ_1^2 as $\sigma_1^{*2} = (n-1)s_1^2/\chi_{n-1}^2$ where χ_{n-1}^2 is chi-square random variable with $n-1$ degrees of freedom and n is the sample size. Generate μ_1 as $\mu_1^* = \bar{y}_1 + Z\sigma_1^*/\sqrt{n}$ where Z is a standard normal deviate.

- Let $\hat{\theta} = (X^T X)^{-1} X^T Y_2$ be the least squares estimate of the regression of Y_2 on Y_1, where X is a $n \times 2$ matrix of a column of ones and the values of y_1 and Y_2 is a $n \times 1$ column vector of values of y_2. Let C be the Cholesky decomposition such that $CC^T = (X^T X)^{-1}$. Let RSS be the residual sum of squares, $(Y_2 - X\hat{\theta})^T (Y_2 - X\hat{\theta})$. Define the draws of (σ_2, θ), as $\sigma_2^{*2} = RSS/\chi_{n-2}^2$ and $\theta^* = \hat{\theta} + \sigma_2^* CZ$ where χ_{n-2}^2 is a chi-square random variable with $n-2$ degrees of freedom and Z is a 2×1 vector of independent standard normal deviates.

- Finally, for (ϕ, σ_3) define Y_3 as a $n \times 1$ vector of values of y_3 and X as a $n \times 4$ matrix with a column vector of ones, values of y_1, y_2 and their product. Use the same procedure as in the previous step. The residual degrees of freedom, however, is $n - 4$.

The predictive distributions needed in Step 3 (or in Step 2a) are a bit complicated. From the joint distribution, focusing on the terms involving y_3, it is easy to see that the predictive distribution for y_3 is given by the regression of y_3 on y_1, y_2 and $y_1 y_2$ (the last term in the exponent). This involves the generated values of (ϕ, σ_3) and the covariate values of y_1 and y_2 for a subject with missing value in y_3.

Focusing on the terms involving y_1 in Equation (1.1), the predictive distribution of y_1 given y_2 and y_3 is proportional to,

$$\exp\left[-\left\{\left(\frac{y_1 - \mu_1}{\sigma_1}\right)^2 + \left(\frac{y_1 - (y_2 - \theta_o)/\theta_1}{\sigma_2/\theta_1}\right)^2 + \right.\right.$$
$$\left.\left.\left(\frac{y_1 - (y_3 - \phi_o - \phi_2 y_2)/(\phi_1 + \phi_3 y_2)}{\sigma_3/(\phi_1 + \phi_3 y_2)}\right)^2\right\}\right],$$

which is a product of three normal densities.

Thus, the predictive distribution is normal with the conditional mean of y_1 given ω, y_2 and y_3 as

$$E(Y_1|y_2, y_3, \omega) = \frac{\mu_1/\sigma_1^2 + \theta_1(y_2 - \theta_o)/\sigma_2^2 + (\phi_1 + \phi_3 y_2)(y_3 - \phi_o - \phi_2 y_2)/\sigma_3^2}{1/\sigma_1^2 + \theta_1^2/\sigma_2^2 + (\phi_1 + \phi_3 y_2)^2/\sigma_3^2}$$

(1.3)

and the conditional variance as

$$Var(Y_1|y_2, y_3, \omega) = (1/\sigma_1^2 + \theta_1^2/\sigma_2^2 + (\phi_1 + \phi_3 y_2)^2/\sigma_3^2)^{-1} \quad (1.4)$$

Similarly, the predictive distribution of y_2 given y_1 and y_3 is proportional to,

$$\exp\left[-\left\{\left(\frac{y_2 - \theta_o - \theta_1 y_1}{\sigma_2}\right)^2 + \left(\frac{y_2 - (y_3 - \phi_o - \phi_1 y_1)/(\phi_2 + \phi_3 y_1)}{\sigma_3/(\phi_1 + \phi_3 y_2)}\right)^2\right\}\right],$$

which is a product of two normal densities. Thus, the predictive distribution of y_2 given y_1, y_3 and ω is normal with mean

$$\frac{(\theta_o + \theta_1 y_1)/\sigma_2^2 + (\phi_1 + \phi_3 y_1)(y_3 - \phi_o - \phi_1 y_1)/\sigma_3^2}{1/\sigma_2^2 + (\phi_1 + \phi_3 y_1)^2/\sigma_3^2}$$

and variance

$$Var(Y_2|y_1, y_3, \omega) = (1/\sigma_2^2 + (\phi_1 + \phi_3 y_1)^2/\sigma_3^2)^{-1}.$$

1.10.3   Comparison of Approaches

How does the joint model approach compare with the empirically derived regression models in the SRMI approach? Note that Equation (1.3) is of the form,

$$\frac{a_o + a_1 y_2 + a_3 y_2^2 + a_4 y_2 y_3}{b_o + b_1 y_2 + b_3 y_2^2},$$

where the coefficients a's and b's are the functions of the parameters. Consider a linear approximation of the equation (1.3) using the Taylor series to obtain,

$$E(Y_1 | y_2, y_3) \approx A_o + A_1 y_2 + A_2 y_3 + A_3 y_2 y_3 + A_4 y_2^2 + A_5 y_2^2 y_3 + \ldots$$

which is similar to the model developed based on the regression analysis. Equation (1.4) indicates a non-constant variance but is not prominent in the regression analysis of the actual data.

Similarly, the linear approximation of the conditional mean of Y_2 given Y_1 and Y_3 is of the form,

$$B_o + B_1 y_1 + B_2 y_3 + B_3 y_1 y_3 + B_4 y_1^2 + + B_5 y_1^2 y_3 + \ldots$$

which is similar to the model used in the SRMI approach. Again, the non-constance variance was not prominent in the regression analysis. Thus, the functional forms of the predictive distributions are similar under the two approaches but the number of parameters fit is different. The SRMI approach uses a total of 17 parameters (14 regression coefficients and 3 residual variances) whereas the joint model approach uses 10 unknown parameters (3 variances and 7 mean parameters). The SRMI approach uses a more flexible system of equations involving the same predictor terms for generating imputations but assumes a constant variance for the residuals. The particular data set did not show much evidence of heteroscedasticity but the SRMI approach can be modified to accommodate non-constant variance (See Chapter 11).

1.10.4   Alternative Modeling Strategies

Development of statistical models for imputation (for that matter, in any statistical analysis) is both an art and science. One strategy is to transform the variables to achieve approximate multivariate normality by considering the Box-Cox transformation,

$$u_j = \frac{(y_j + c_j)^{\lambda_j} - 1}{\lambda_j},$$

where $j = 1, 2, 3$ and c_j is constant to make all the values positive. $\lambda_j, j = 1, 2, 3$ are such that (u_1, u_2, u_3) is approximately multivariate normal with mean μ and covariance matrix Σ. The missing values can be imputed on the transformed scale and re-transformed to the original scale.

Sometimes one gets trapped in the notion of a "true" or "correct" model.

This notion is useful while developing and evaluating statistical procedures (such as estimators, significance tests, confidence intervals etc.) from the repeated sampling perspective. As a practitioner, it is important to remember that "All models are wrong but some are useful" (this quote is often attributed to a great statistician, George Box). The models are an approximate description of a plausible probabilistic mechanism generating the observed data, and thus allowing the development of interpretable scientific quantities or estimands of interest and statistical inferences about them. It is important, therefore, to be careful in developing models by making sure that they fit the data well, they are substantively sensible and one should not overly interpret the posited models as the functioning of the nature. In other words, the models are useful ways of constructing data summaries or statistics that enable us to study the target population of interest.

If one entertains several competing models, then sensitivity of inferences to model assumptions should be carefully evaluated. The goal of building the imputation model is to use all available observed information, empirical investigations and scientific understanding to develop the best possible approach for predicting the missing values and the associated uncertainties in the prediction.

1.11   Complex Sample Surveys

A finite population sample survey consists of selecting a sample (using a probabilistic sampling process or mechanism) from a well defined list, a sampling frame, and then administering a questionnaire, or other means, for collecting data. A probability sample requires that every individual in the sampling frame has a positive probability of being included in the sample.

Often to capture important features of the population (information in the sampling frame), the sampling mechanism uses stratification of the population and unequal probabilities of selection to achieve specific objectives. To reduce the cost of administration (for example, interviewer travel costs), the sample will be selected in clusters and then clusters may be sub-sampled for individual units. These complex issues introduce stratification, weighting and clustering as standard features in many, if not all, sample surveys.

Analysis of data from these surveys, even without any missing data, need to incorporate these features. A variety of approaches are available to incorporate them in the complete data analysis and implemented in many software packages. For example, *IVEware* uses a linearization or Jackknife Repeated Replication technique to incorporate these features when constructing inferences.

The imputation process also need to include these features when creating imputation of the missing values. There are three possible approaches depending upon the level of correlation between the design variables and survey variables with missing values, and the fraction of missing cases. If the fraction of missing cases is small and the design variables are weakly correlated with the survey variables with missing values then one can ignore the design variables during the imputation but incorporate them in the completed data analysis.

The second approach is to include strata as dummy variables (or use the variables used in the construction of strata as predictors), survey weight as a predictor and cluster level random effects to account for correlation between observations within the same cluster. Instead of random effects for clusters, cluster level covariates may be used as predictors.

The number of design variables can be large in some surveys. Therefore, one may want to use some dimensionality reduction techniques to create summary scores. As an example suppose that Z denotes a vector of design variables and R denotes the response indicator taking the value 1, if the variable under consideration is observed and 0 if the variable is missing. A well fitting logistic regression model,

$$Pr(R = 1|Z) = [1 + \exp(-Z^T\gamma)]^{-1}$$

may be fit and the estimated linear predictor $Z^T\widehat{\gamma}$, where $\widehat{\gamma}$ is the maximum likelihood estimate of γ, may be considered as a scalar summary. This balancing score may be included as an additional predictor or covariate in the sequential regression imputation model for the variable under consideration.

The third approach is to reconstruct synthetic populations based on the design variables using a non-parametric Bayes approach and then multiply impute the missing values in the synthetic populations. This may be viewed as a four step process: (1) "Un-complex" the sampling unit selection process through creation of several synthetic populations; (2) Multiply impute the missing values in the variables for all the sampled units propagated in each synthetic population; (3) Compute the population quantity from each imputed synthetic population; and (4) combine all the computed quantities to construct inferences. This approach is computationally intensive and, perhaps, the most principled approach of the three methods described here. Chapter 9 discusses this method in more details.

1.12   Imputation Diagnostics

It is very important to check whether the imputations created under a set of model assumptions are sensible and comparable to the observed set of values. There are two possible approaches for checking the imputations. The first approach is to compare observed and imputed values for any specific variable conditional on the observed values of other variables. The second approach is to generate synthetic data sets (all imputed values under the stated imputation model assumptions) and compare to the observed data sets. Both approaches are possible using *IVEware*.

1.12.1   Propensity Based Comparison

Let y_j be the variable of interest and $y_{j,obs}$ and $y_{j,imp}$ be the observed and imputed values, respectively, for this variable. Let $y_{obs}^{(-j)}$ denote the observed values in all other variables other than y_j. Imputations from a well fitting model should satisfy,

$$pr(y_{j,obs}|y_{obs}^{(-j)}) \approx pr(y_{j,imp}|y_{obs}^{(-j)}). \tag{1.5}$$

That is, conditional on $y_{obs}^{(-j)}$, the observed and imputed values should be similar in distribution. One possible approach to assess Equation (1.5) is through the propensity score method.

Let $R_j = 1$ for the observed values in y_j and 0 for the imputed values. Let $\widehat{p}_j = Pr(R_j = 1|y_{obs}^{(-j)})$ be the estimated propensity score for the observed versus imputed values in y_j. Equation (1.5) can be restated as

$$pr(y_{j,obs}|\widehat{p}_j) \approx pr(y_{j,imp}|\widehat{p}_j)$$

IVEware estimates the propensity scores and compares the distributions of imputed and observed values conditional on the propensity score in a number of ways. One approach for comparison of distributions is through a scatter plot of the vector y_j^* (containing observed and imputed values) against p_j with different color or symbols for the observed and imputed values. In this scatter plot, the observed and imputed values should appear exchangeable conditional on p_j. Another approach is to compare the distribution of the residuals of y_j^* on p_j for $R_j = 0$ and $R_j = 1$. For example, the histograms or the kernel densities of the residuals should be similar for the observed ($R_j = 1$) and imputed values ($R_j = 0$). This approach will be illustrated through examples in the later chapters.

1.12.2   Synthetic Data Approach

The suite of regression models used in the imputation of all the variables can be checked by creating synthetic data sets and then comparing the synthetic data sets with the observed data set. The idea is that, if the model is reasonable then the observed data sets should be within the realm of all these synthetic data sets. The **SYNTHESIZE** module in *IVEware* can be used to generate several synthetic data sets as illustrated in Figure 1.3. The observed values in the original data can be compared to the corresponding values in the synthetic data sets to assess the model assumptions. This approach is similar to posterior predictive checks in a standard Bayesian analysis.

Figure 1.3: A schematic display of synthetic data sets produced by **SYN-THESIZE** module in *IVEware*.

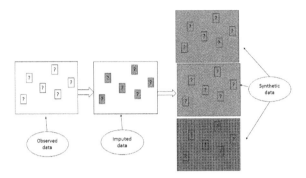

The challenge is to develop a metric to compare a data set with another data set. The box plots of the variables, probability-probability or Quantile-Quantile plots, key statistics such mean, variances, and correlation coefficients are a few approaches that can be used to compare the observed and synthetic data sets.

1.13   Should We Impute or Not?

Suppose that a predictor variable X in a multiple linear regression model (where other covariates also have some small percentage of missing values) has a lot of missing values (say, 75% of the sample). A question arises, "Should I impute so many missing values?" In a response to this question, it is better to ask the second question, "What is the alternative to imputation?", before trying to answer the first question. If the answer to the second question is "complete-case analysis" then obviously it does not make sense because more than 75% of subjects will be excluded from the analysis with a potential for bias and a definite loss of efficiency.

If the answer to the second question is "Drop the variables from the analysis" then obviously no imputation is necessary. However, the substantive analysis has now been modified by dropping the variable and, consequently, a different scientific question is being answered in the new analysis.

As long as this variable is important for the analysis, it is better to impute the missing values using as much information as possible. If one expects a large amount of missing values in a particular variable, it is better to consider auxiliary predictors of this variable, collect them and include them in the imputation process.

Suppose that income is a variable which is expected to have a substantial number of missing values. It is better to collect correlates of income, such as house value (from the administrative data), the interviewer observations about the house and neighborhood, the number, model and make of the automobiles etc, which may have good predictive power for predicting the missing income, in addition to the traditional measures, demographics, occupation, employment, industry etc. Thus, the planning for adjustment for missing data should begin at the design stage of the study, not as post-hoc thinking.

1.14   Is Imputation Making Up Data?

This question commonly arises among many researchers based on superficial reading or understanding of the imputation based approach as the method for dealing with missing data. The answer is yes, if one were to impute the missing values and then **pretend as though we have complete data**. The view should be: *Impute the missing values but analyze the imputed data accounting for the imputation.* That is, estimation of uncertainty about the estimates of (or inference about) the target quantity of interest must reflect the fact that some are real (observed values) and others (imputations) are "made up" from the real.

Consider an extreme (or rather a silly) example. Suppose a survey was planned with a sample of size 1,000 to collect information on a continuous variable Y. However, only 100 subjects responded, yielding the respondent mean of Y values as 25 and a standard deviation of 5. Assume a MCAR mechanism. The correct standard error of the sample mean is obviously $5/\sqrt{100} = 0.5$. Now, suppose that the 900 missing values were imputed using the approximate Bayesian Bootstrap method described Section 1.5 (by redrawing 900 values from 100 values). It so happens that the mean and the standard deviation from the filled-in or completed data are 25 and 5, respectively. If the completed data of size 1,000 were to be analyzed naively, the standard error output will be $5/\sqrt{1000} \approx 0.16$. Obviously, this is incorrect because the "information content" of the completed data is the same or even less (because of adding noise) than the observed data. Therefore, the standard error of the mean has to be larger than or equal to 0.5.

Multiple imputation is one approach for incorporating the uncertainty about the imputed values. The goal of the multiple imputation based approach should be viewed as the "rectangularization" of most, if not all, of the information in the observed data set such that complete data software can be used to perform a proper analysis that reflects the uncertainty in the imputations.

1.15   Multiple Imputation Analysis

1.15.1   Point and Interval Estimates

As indicated earlier, any analysis of imputed data must incorporate the imputation uncertainty. Multiple imputation is one method for incorporating this uncertainty. In this approach, a chosen imputation method is repeated, say, M times, to generate M completed data sets. Each completed data set is analyzed separately to obtain quantities of interest. Suppose that e_l is the estimate of a parameter of interest from the completed data set $l = 1, 2, \ldots, M$ and the corresponding sampling variance is U_l. The average, $\bar{e}_{MI} = \sum_l e_l / M$ is the multiple imputation estimate of the parameter. The average $\bar{U}_{MI} = \sum_l U_l / M$ is approximately the variance that would have been obtained had there been no missing data. The additional uncertainty due to missing values (imputation uncertainty) is measured as $B_{MI} = (1 + 1/M)B_M$ where

$$B_M = \sum_l (e_l - \bar{e}_{MI})^2 / (M - 1).$$

The multiple imputation variance estimate is $T_{MI} = \bar{U}_{MI} + B_{MI}$. The quantity $r_{MI} = B_{MI}/T_{MI}$ is the fraction of missing information, the proportion the total variance that is due to missing data. Another quantity of interest is $g_{MI} = B_{MI}/\bar{U}_{MI}$, the proportionate increase in variance due to missing values relative to an estimated complete-data variance, $\bar{U}_{MI}$.

For constructing confidence intervals, a t distribution is used with $\nu_{MI} = (M - 1)/r_{MI}^2$ degrees of freedom. This degrees of freedom assumes that the complete data analysis is infinite (or large sample size). If the complete data analysis is small then further adjustment is needed. Let ν be the complete data degrees of freedom. Defining

$$C_{MI} = \frac{\nu + 1}{\nu + 3}\nu(1 - r_{MI}),$$

the adjusted degrees of freedom is

$$\nu_{MI}^* = \frac{\nu_{MI}C_{MI}}{\nu_{MI} + C_{MI}}.$$

1.15.2   Multivariate Hypothesis Tests

Let θ be a p-dimensional vector and the inference about θ involves testing the null hypothesis $H_o : \theta = \theta_o$. As before, let $e_1, e_2, \ldots, e_M$ be the M completed data sets and $U_l, l = 1, 2, \ldots, M$ be the corresponding $p \times p$ variance-covariance matrices. Define $\bar{e}_{MI} = \sum_l e_l / M$, $\bar{U}_{MI} = \sum_l U_l / M$ and $B_{MI} = (1 + 1/M) \sum_l (e_l - \bar{e}_{MI})(e_l - \bar{e}_{MI})^T / (M - 1)$. The multiple imputation variance-covariance matrix is $T_{MI} = \bar{U}_{MI} + B_{MI}$. When M, the number of imputations is small relative to p, the matric B_{MI} may not be non-singular and so T_{MI} can be quite unstable. If the effect of missing data is similar across the p, parameters, as measured by the fraction of missing information, then one can approximate $T_{MI} \approx \tilde{T}_{MI} = (1 + g_{MI})\bar{U}_{MI}$ where g_{MI} is the average trace of the matrix, $G_{MI} = B_{MI}\bar{U}_{MI}^{-1}$ (the trace of a matrix is the sum of its diagonal elements and, hence $g_{MI} = Tr(G_{MI})/p$).

Define the test statistic,

$$D_{MI} = (\bar{e}_{MI} - \theta_o)^T \bar{U}_{MI}^{-1} (\bar{e}_{MI} - \theta_o)/[p(1 + g_{MI})].$$

An approximate sampling distribution of D_{MI} is an F-distribution with p as the numerator degrees of freedom and the denominator degrees of freedom,

$$w_{MI} = 4(t - 4)[1 + (1 - 2t^{-1})/g_{MI}]^2$$

where $t = p(M - 1)$ is assumed to be greater than 4. If $t \le 4$ then $w_{MI} = t(1 + p)(1 + 1/g_{MI})^2/2$. The derivation of w_{MI} is based on asymptotic distributions (that is, assumes large samples). If the sample size is small (that is, if the complete-data degrees of freedom is, ν_C), the modification of w_{MI} is

$$w_{MI}^* = \frac{w_{MI}c_M}{w_{MI} + c_M}$$

where

$$c_M = \frac{(\nu_C + 1)\nu_C}{(\nu_C + 3)(1 + g_{MI})}.$$

An alternative, more complicated, denominator degrees of freedom derived in Reiter (2007) is

$$w_{MI}^* = 4 + \left[\frac{1}{\nu_C^* - 4(1 + a_{MI})} + \frac{1}{t - 4} \left(\frac{a_{MI}^2(\nu_C^* - 2(1 + a_{MI}))}{(1 + a_{MI}^2)(\nu_C^* - 4(1 + a_{MI}))} \right) \right]^{-1}$$

where $\nu_C^* = \nu_C(\nu_c + 1)(\nu_C + 3)^{-1}$ and $a_{MI} = tg_{MI}/(t - 2)$.

In some software packages, it is difficult to compute g_{MI}. The fraction of missing information, however, for the p parameters, $r_{j,MI}, j = 1, 2, \ldots, p$, is provided as a part of the output. An approximation, $1 + g_{MI} \approx \bar{r}_{+,MI}/(1 - \bar{r}+, MI)$, may be used, where $\bar{r}_{+,MI} = \sum_j r_{j,MI}/p$, is the average fraction of missing information. If the fraction of information for the p parameters is not equal, it may be prudent to use the largest fraction of missing information to calculate the denominator degrees of freedom.

1.15.3 Combining Test Statistics

In many completed data analyses, such as goodness of fit tests, the likelihood ratio tests, model comparisons using the generalized estimating equations, log-linear model, contingency table analysis etc., involves computing chi-square statistics. One may then need a framework for combining these test statistics across the multiply imputed data sets. Let $d_1, d_2, \ldots, d_M$ be the M completed-data chi-square test statistics and let the completed-data degrees of freedom (or the number of parameters)be k. Let $\bar{d}_{MI} = \sum_l d_l/M$ be their average. Define $\bar{p}_{MI} = \sum_l \sqrt{d_l}/M$ and $v_{MI} = (1 + 1/M)\sum_l(\sqrt{d_l} - \bar{p}_{MI})^2/(M - 1)$. The combined test statistic is defined as,

$$\tilde{D}_{MI} = \frac{\bar{d}_{MI}/k - (M + 1)v_{MI}/(M - 1)}{1 + v_{MI}},$$

and is referred to an F distribution with k as the numerator degrees of freedom and,

$$\nu_d = k^{-3/M}(1 + 1/v_{MI}^2),$$

as the denominator degrees of freedom.

There are many other combining rules and these will be illustrated in the later chapters.

1.16   Multiple Imputation Theory

Theoretical justification for the combining rules involves both repeated sampling and Bayesian perspectives. The notion that information about any unknown quantity needs to be expressed through a probability distribution over the possible values of that quantity is fundamental to this theoretical justification. Suppose that T_n is an observable quantity that could be constructed based on a sample of size n . The statement that $(T_n - \mu)/\sqrt{V}|\mu, V \sim N(0,1)$ expresses uncertainty about the plausible values of T_n for a given values of μ and V. The statement $(T_n - \mu)/\sqrt{V}|T_n, V \sim N(0,1)$ expresses uncertainty about the plausible values of μ for a given value of T_n and V. The basis of the first normal distribution is the repeated sampling of samples of size n given the population values of μ and V and the basis of the second normal distribution is the Bayesian posterior distribution of μ.

The same notion can be extended when T_n cannot be analytically computed but is approximated using a procedure, yielding $\tilde{T}_n$. The statements $(T_n - \tilde{T}_n)/\sqrt{U}|T_n, U \sim N(0,1)$ and $(T_n - \tilde{T}_n)/\sqrt{U}|U, \tilde{T}_n \sim N(0,1)$ are both valid to, respectively, express the uncertainty about the accuracy of the approximation procedure and the predictive uncertainty about actual quantity of interest, T_n given the yield from the procedure $\tilde{T}_n$. An example is a numerical or Monte Carlo technique for approximating T_n where the value of U may be chosen so that the difference between T_n and $\tilde{T}_n$ can be made arbitrarily small. In such cases, the notation, $(T_n - \tilde{T}_n)|\tilde{T}_n \sim N(0, << \tilde{T}_n)$, is used to indicate that the predictive uncertainty in T_n is very small when compared to $\tilde{T}_n$. It is important to recognize that, conceptually, T_n is still an unknown quantity (if it cannot be calculated using an analytical formula) even if all the sample values are known. There are situations where U cannot be made arbitrarily small given the cost or computational constraints and this additional uncertainty should be incorporated when constructing inferences. Consider now the justification of multiple imputation with this understanding of the sources, and their expressions, of uncertainty through probability distributions.

Let $f(y|\theta)$ be the substantive model of interest. The goal is to infer about the scalar parameter θ. With the fully observed data, $D = \{y_i, i = 1, 2, \ldots, n\}$, on n subjects, suppose that $\widehat{\theta}_D$ is the most efficient estimate of θ and its sampling variance be $U = u(\theta)$. Furthermore, assume that the sample size is large enough to permit the asymptotic approximation,

$$U^{-1/2}(\theta - \widehat{\theta}_D)|\theta, U \sim N(0,1).$$

Let U_D be an estimate of U such that $U_D/U|U \sim N(1, << U)$. These approximations are to be interpreted from the repeated sampling perspective and is an expression about the uncertainty in the procedure used to construct $\widehat{\theta}_D$ to estimate θ.

From the Bayesian perspective, assume $U_D^{-1/2}(\widehat{\theta}_D - \theta)|U_D, \widehat{\theta}_D \sim N(0,1)$ and $U_D/U|U_D \sim (1, << U_D)$, a large sample approximation of the posterior distribution of θ and U, conditional on the data D. It is important to note the quantities subject to uncertainty (quantities before the |) and what is being conditioned on(quantities after the |) in the expressions of uncertainty.

With missing data, some elements of D are not known. Let y_{obs} denotes the observed portion of D and y_{mis} denotes its missing portion. Assume that missing data mechanism is ignorable, and thus an explicit formulation of the distribution for the response indicator, R, is not needed. Thus, now both $\widehat{\theta}_D$ and U_D are the functions of the known (y_{obs}) and the unknown (y_{mis}) values. Let $g(y_{mis}|y_{obs})$ be the predictive (imputation) distribution of the missing values, y_{mis}, given y_{obs}. Let

$$\widehat{\theta}_o = E(\widehat{\theta}_D|y_{obs}) = \int \widehat{\theta}_D g(y_{mis}|y_{obs}) dy_{mis}$$

be the average of the complete data estimate with respect to the imputation distribution and

$$U_o = E(U_D|y_{obs}) = \int U_D(y_{mis}|y_{obs}) dy_{mis}.$$

Thus,

$$Var(\widehat{\theta}_D|y_{obs}) = E(U_D|y_{obs}) + Var(E(\widehat{\theta}_D)|y_{obs}) = U_o + B_o$$

where B_o is the increase in variance due to missing values. Furthermore, assume that the posterior distribution of θ, conditional on (θ_o, U_o, B_o), may be approximated as $N(\theta_o, U_o + B_o)$. The quantity, $r_o = B_o/(U_o + B_o)$, is the fractional increase in variance due to missing values or the fraction of missing information attributed to missing data.

Assume that $\widehat{\theta}_o, U_o$ and B_o, cannot be analytically calculated and a Monte Carlo procedure is to be used through drawing values from the predictive or imputation distribution, $g(y_{mis}|y_{obs})$. Let $y_{mis}^{(l)}, l = 1, 2, \ldots$ be the draws from the predictive distribution $g(y_{mis}|y_{obs})$ and let $\widehat{\theta}_l$ and U_l be the estimate and its variance computed from the completed data, $D^{(l)} = (y_{obs}, y_{mis}^{(l)})$. These are the estimate and its variance derived from the complete procedure (that is, the procedure that would have been used to compute $(\widehat{\theta}_D, U_D)$ with a complete data) .

The multiple imputation estimate is $\bar{\theta}_{MI} = \sum_l \widehat{\theta}_l/M$ and let $B_M = \sum_l (\widehat{\theta}_l - \bar{\theta}_{MI})^2/(M-1)$, be the variance of the completed data estimates across the M imputations. Let $\bar{\theta}_\infty = lim_{M\to\infty}\bar{\theta}_{MI}$ and $B_\infty = lim_{M\to\infty}B_M$ be the limiting quantities as the number of imputations tends to ∞. Assume $\bar{U}_{MI} = \sum_l U_l/M$ tends to $\bar{U}_\infty$ as $M \to \infty$.

Assume that the imputation procedure is such that $\bar{\theta}_\infty = \widehat{\theta}_o$, $B_\infty = B_o$ and $U_\infty = U_o$. That is, the observed data quantities are recoverable from the infinite number of imputations under the imputation model.

When the number of imputations is small, assume that the uncertainty in the approximation procedure, due to finite M, is captured by

S1. $B_\infty^{-1/2}(\bar{\theta}_{MI} - \bar{\theta}_\infty)/\sqrt{M}|\bar{\theta}_\infty, B_\infty \sim N(0,1)$

S2. $(M-1)B_M/B_\infty|B_\infty \sim \chi^2_{M-1}$

S3. $\bar{U}_{MI}/U_\infty|U_\infty \sim (1, << U_\infty)$

As discussed earlier, the three items enumerated above could be transformed as an expression of uncertainty about $(\theta_\infty, B_\infty, U_\infty)$, conditional on $(\bar{\theta}_{MI}, B_{MI}, \bar{U}_{MI})$ as follows:

P1. $\theta_\infty|\bar{\theta}_{MI}, B_\infty \sim N(\bar{\theta}_{MI}, B_\infty/M)$

P2. $B_\infty|B_M \sim (M-1)B_M/\chi^2_{M-1}$

P3. $U_\infty|\bar{U}_{MI} \sim (\bar{U}_{MI}, << \bar{U}_{MI})$

Thus, now the ultimate goal of constructing the posterior distribution of θ conditional on the approximate quantities $(\bar{\theta}_{MI}, B_M, \bar{U}_{MI})$, reduces to the following expressions:

U1. $\theta|\bar{\theta}_\infty, B_\infty, \bar{U}_\infty \sim N(\theta_\infty, \bar{U}_\infty + B_\infty)$ (Follows from $\theta|\theta_o, U_o, B_o \sim N(\theta_o, U_o + B_o)$ and $\theta_o = \bar{\theta}_\infty, \bar{U}_\infty = U_o$ and $B_o = B_\infty$)

U2. $\theta|\bar{\theta}_\infty, B_\infty, \bar{U}_{MI} \sim N(\bar{\theta}_\infty, \bar{U}_{MI} + B_\infty)$ (replacing $\bar{U}_\infty$ by $\bar{U}_{MI}$, permitted under (P3) above)

U3. $\theta|\bar{\theta}_{MI}, \bar{U}_{MI}, B_\infty \sim N(\bar{\theta}_{MI}, \bar{U}_{MI} + (1+1/M)B_\infty)$ (integrating with respect to $\bar{\theta}_\infty$, permitted under (P1))

The final step is to integrate with respect to the posterior distribution of B_∞, conditional on B_M to arrive at

$$Pr(\theta|\bar{\theta}_{MI}, \bar{U}_{MI}, B_M) = \int \phi \left(\frac{\theta - \bar{\theta}_{MI}}{\sqrt{\bar{U}_{MI} + (1 + 1/M)B_\infty}} \right) \times$$

$$f(B_\infty|\bar{\theta}_{MI}, \bar{U}_{MI}, B_M)dB_\infty, \qquad (1.6)$$

where $\phi()$ is the normal density function. There are several possible approximations of this integral but consider a t approximation for analytical convenience, and this desire for a t approximation, necessitates approximating the posterior distribution of $[\bar{U}_{MI} + (1 + 1/M)B_\infty]^{-1}$, conditional on $(\bar{\theta}_{MI}, \bar{U}_{MI}, B_M)$, by a scaled chi-square distribution $a\chi_\nu^2$.

Given (P2),

$$[\bar{U}_{MI} + (1 + 1/M)B_\infty]^{-1} \equiv [\bar{U}_{MI} + (1 + 1/M)B_M X^{-1}]^{-1}$$

where X has a scaled chi-square distribution, $\chi_{M-1}^2/(M-1)$. To derive the expressions for a and ν, the mean and variance of $[\bar{U}_{MI}+1(1+1/M)B_\infty]^{-1}$ will be equated to the mean $(a\nu)$ and the variance $(2a\nu^2)$ of the scaled chi-square distribution $a\chi_\nu^2$.

Taylor's expansion, around the mean of X (=1) yields,

$$[\bar{U}_{MI} + (1 + 1/M)B_M X^{-1}]^{-1} \approx [1 + r_{MI}(X - 1)]/T_{MI}$$

where $T_{MI} = \bar{U}_{MI} + (1 + 1/M)B_M = \bar{U}_{MI} + B_{MI}$ and $r_{MI} = (1 + 1/M)B_M/T_{MI} = B_{MI}/T_{MI}$, an estimate of the fraction of information. It is easy to show that

$$a\nu = 1/T_{MI},$$

and

$$2a^2\nu = 2r_{MI}^2/[(M-1)T_{MI}^2],$$

leading to $a = r_{MI}^2/[(M-1)T_{MI}] = 1/(\nu T_{MI})$ where $\nu = (M-1)/r_{MI}^2$. Thus,

$$\frac{T_{MI}}{\bar{U}_{MI} + (1 + 1/M)B\infty} \sim \frac{\chi_\nu^2}{\nu}$$

yielding

$$\theta|\bar{\theta}_{MI}, \bar{U}_{MI}, B_{MI} \sim t_\nu(\bar{\theta}_{MI}, T_{MI})$$

a t distribution with location $\bar{\theta}_{MI}$, scale T_{MI} and the degrees of freedom $\nu = \nu_{MI} = (M-1)/r_{MI}^2$.

1.17   Number of Imputations

Multiple imputation is a simulation approach for constructing the inferences (for example, confidence intervals). The degrees of freedom for constructing the confidence interval, $\nu_{MI} = (M-1)/r_{MI}^2$, depends upon the number of imputations M and the fraction of missing information r_{MI}.

Obviously, M has to be greater than or equal to 2, otherwise B_{MI} cannot be calculated. For B_{MI} (and therefore r_{MI}) to be reliably estimated we need M to be large. Also, M should be large for the degrees of freedom ν_{MI} to be large (for shorter confidence intervals). For $\nu_{MI} \geq 50$, one needs $M \geq 50r_{MI}^2 + 1$. For example, with 25% as the fraction of missing information, one needs $M \geq 50/16+1 \geq 4.125$, hence $M = 5$ was suggested initially (See Rubin (1977)). This number was also proposed in an era where the computational landscape was quite different, both in terms of storage and the data processing power. Nevertheless, $M = 5$ may not produce a reliable estimate of r_{MI} and, therefore, ν_{MI} is subject to considerable sampling variability.

The uncertainty in B_{MI} is captured through it is distribution, conditional on the observed values, given by $(M-1)B_M/B_\infty \sim \chi_{M-1}^2$ (See (S2) or (P2) in the previous section). Thus, assuming that $\bar{U}_M \approx \bar{U}_\infty$, obtains,

$$r_{MI} = \frac{(1+1/M)B_\infty \times X}{\bar{U}_\infty + (1+1/M)B_\infty \times X}$$

where $X = \chi_{M-1}^2/(M-1)$. Taylor's expansion of the expression for r_{MI} as a function of X around its mean of 1, results in

$$r_{MI} \approx r_\infty + (X-1)r_\infty(1-r_\infty),$$

where $r_\infty = (1+1/M)B_\infty/[(1+1/M)B_\infty + \bar{U}_\infty]$

Thus, the uncertainty in r_{MI} can be expressed as

$$var(r_{MI}) \approx 2r_\infty^2(1-r_\infty)^2/(M-1)$$

which has the maximum of $1/[8(M-1)]$ at $r_\infty = 1/2$. Thus, the coefficient of variation of, r_{MI}, at the maximum is $1/\sqrt{2(M-1)}$. For this coefficient of variation to be less than c, one needs $M \geq 1 + 1/(2c^2)$. For example, with $c = 0.25$, M should be at least 9, with $c = 0.1$, M should be at least 51 and for $c = 0.05$, M should be at least 201.

Current computational resources are such that hundreds and thousands of bootstrap samples and Monte Carlo simulations can be performed in a flash. Thus, choosing a large value of M should not be a burden. It is possible to choose M adaptively. For example, choose, say, $M = 10$ and compute the fraction of missing information and then use the largest fraction of information to determine the number of additional imputations needed.

1.18   Additional Reading

The basic textbooks for statistical background are Hogg, McKean and Craig (2012), Casella and Berger (2002), Cox and Hinkley (1979) and a two volume book by Bickel and Doksum (2006). For a more in-depth understanding of Bayesian analysis, refer to Box and Tiao (1973), Carlin and Louis (2008) and Gelman et al (2013). There are numerous other excellent text books but the previously mentioned books provide examples of the full spectrum of the field and/or the basis of statistical inference. For background on complex surveys, see the classic texts by Cochran (1977) and Kish (1995). Also, Lohr (2009), Thompson (2012) and Valliant, Dever and Kreuter (2013) provide more recent and excellent coverage of the relevant topics.

As indicated in the preface, the objective of this book is to provide sufficient information to perform multiple imputation analysis for a variety of statistical models. The software *IVEware* is used for illustrative purposes. In general, these details are enough to implement these methods using other software packages such as "mi" or 'ice" in Stata (Royston (2007), STATACorp(2017)), "PROC MI/PROC MIANALYZE" in SAS (Berglund and Heeringa (2014)), or MICE in R (Van Buuren (2012)). However, for a conceptual understanding of issues of missing data analysis, refer to the classic text Little and Rubin (2002). Raghunathan (2016) covers the weighting and multiple imputation methods from a practical perspective. A classic text book providing detailed knowledge about multiple imputation is Rubin (1987) and a more recent textbook is Carpenter and Kenward (2014). Additional general books on missing data include Allison (2002), Molenberghs and Kenward(2007) and Schafer (1997).

An excellent reference for Tukey's *gh* distribution is Hoaglin, Mosteller, Tukey (1985). This distribution was first used for imputation in He and Raghunathan (2006), and generalized in He and Raghunathan (2012). The nonparametric approach discussed in Section 1.5 is from a technical report by Bondarenko and Raghunathan (2010).

There are excellent texts discussing model building strategies such as Gelman and Hill (2006), Weisberg (2013), Draper and Smith (1998), Neter, Kutner, Nachtsheim and Wasserman (1996) and Hosmer, Lemeshow and Sturdivant (2013). Imputation diagnostics are discussed in Abayomi, Gelman and Levy (2008) and Bondarenko and Raghunathan (2016). Generation of synthetic data sets is discussed in Raghunathan, Reiter and Rubin (2003), though in a different context. Gelman, Carlin, Stern, Dunson, Vehtari and Rubin (2013) provide ideas about the posterior predictive checks that can be adapted for assessing reasonableness of the imputations.

Rubin (1976a) is the foundation for the missing data mechanism that has changed the view on assessing the appropriate methods for analyzing data with some missing values. Multiple imputation was proposed by Rubin (1978) and most theoretical and practical considerations are described in the classic book Rubin (1987). The theoretical justification given in this chapter is derived in Rubin (1987) and Raghunathan (1987). Rubin and Schenker (1986) develops the combining rules for a scalar parameter. The multiparameter case is developed in Raghunathan (1987) and refined in Li et. al. (1991a). Combining chi-square statistics was also developed in Raghunathan (1987) and extended in Li et. al. (1991b). Barnard and Rubin (1999) developed the modification for small sample degrees of freedom. Reiter (2007) further refined the combining rules. Meng and Rubin (1992) developed a method for performing the likelihood ratio test with multiply imputed data sets and is not considered in the book because it has not been implemented in any software packages. Räghunathan, Solenberger, Berglund, Van Hoewyk (2017) implemented many updates to the software including imputation diagnostics, linear regression diagnostics, finite weighted Bayesian Bootstrap (Zhou (2014) and Zhou, Elliot and Raghunathan, 2016a, 2016b, 2016c) and synthesis of data for confidentiality reasons (Raghunathan, Reiter and Rubin (2003)).

The sequential regression approach was first suggested by Kennickell (1991) for continuous variables in the survey of consumer finances. Brand (1999)in a doctoral dissertation develops this method further and names it as variabe-by-variable imputation. Van Buuren and Oudshoorn (1999) in a technical report introduced a version and named it as Multivariate Imputation by Chained Equations and developed the R package software, MICE. Raghunathan et al (2001) develops the methods for several types of variables, incorporates bounds and restrictions and termed it as Sequential Regression Multivariate Imputation (SRMI). The software *IVEware* implementing this approach is available at www.iveware.org. See also Van Buuren (2007) and White Royston and Wood (2011). There are excellent books on multiple imputation methods that can be used as a companion to the material presented in this book: Little and Rubin (2002), Van Buuren (2012)and Carpenter and Kenward (2013). Many data sets used in this book are also discussed in Raghunathan (2016).

The number of imputations needed for stable inferences are considered in Graham et al (2007) and Allison (2012).

2

Descriptive Statistics

2.1 Introduction

Any analysis of missing data using the multiple imputation approach involves three steps: (1) Imputation; (2) Completed Data Analysis of each imputed data; and (3) Combining statistics across imputations. Imputation may use many variables in the process even though the analysis may involve only a subset of these variables. In the bivariate example discussed in Chapter 1, the imputation may involve both variables Y_1 and Y_2 although the analysis may involve only Y_2 (such as computing the mean or standard deviation). Though, valid inferences can be obtained by just using Y_2 in both the imputation and analysis steps, much efficiency may be gained by borrowing strength from the observed Y_1 to impute Y_2. This chapter focuses on univariate descriptive statistics and some simple comparisons. Hence, the imputation process for the entire set of variables is described first but the subsequent analysis involves only a subset of variables.

Suppose that the population mean of a variable, Y, is $E(Y) = \mu$ with the population variance $\sigma^2 = Var(Y)$. If Y is binary then μ is the population proportion or $Pr(Y = 1) = \mu = E(Y)$ and $\sigma^2 = \mu(1 - \mu)$. The goal is to estimate or infer about μ and σ^2 (or just the μ in the binary case). Normally, the standard deviation, σ, is often reported as it is in the same units as μ. The mean, proportion, and standard deviation are often used to the describe the study population in any resultant research articles or reports.

Consider the continuous variable case and a simple random sample. Let $\bar{y}_l$ and s_l^2 be the sample mean and variance from the completed data $l = 1, 2, \ldots, M$. Let n be the sample size. Sometimes the sample size may not be the same across the completed data sets. Suppose that the population mean of interest is for a subdomain (Age 65 or above, for example) and the variables (Age in the example) defining the subdomain have some missing values and have been imputed (Age, for example). In this situation, n_l will be sample size for subdomain in completed data set $l = 1, 2, \ldots, M$ and the formulas given below should be modified accordingly.

The combining rule is rather simple where $\hat{\mu}_{MI} = \sum_l \bar{y}_l / M$, $\bar{U}_{MI} = \sum_l (s_l^2/n)/M$ (or $\sum_l (s_l^2/n_l)/M$) and $B_{MI} = (1 + M^{-1}) \sum_l (\bar{y}_l - \hat{\mu}_{MI})^2/(M-1)$. The degrees of freedom, assuming n is large, is $\nu_{MI} = (M-1)/r_{MI}^2$ where $r_{MI} = B_{MI}/(\bar{U}_{MI} + B_{MI})$. The confidence interval, with level $100(1 - \alpha)\%$, is $(\widehat{\mu}_{MI} \pm t_{\nu_{MI},\alpha/2}\sqrt{T_{MI}})$ where $T_{MI} = \bar{U}_{MI} + B_{MI}$. The standard error of the estimate $\widehat{\mu}_{MI}$ is $\sqrt{T_{MI}}$.

For inference about the sample variance, note that $\widehat{\sigma}^2_{MI} = \sum_l s_l^2/M$. The multiple imputation standard deviation is the square root, $\widehat{\sigma}_{MI}$. Often the variance of the sample variance is not of interest. But it can be calculated easily. Note that $U_l = Var(s_l^2) = a_l/n - s_l^4(n-3)/(n(n-1))$ where a_l is the average of the fourth power of deviations of the observations from the mean, $\sum_i (y_i - \bar{y})^4/(n-1)$ calculated from the completed data $l = 1, 2, \ldots, M$. If Y is approximately normal then $Var(s_l^2)$ reduces to $2s_l^4/(n-1)$. The standard combining rules can be used to obtain the multiple imputation variance T_{MI} and the fraction of missing information r_{MI}.

A normal approximation for the posterior distribution of $\log \sigma$ (or the sampling distribution of $\log s_l$) might be more reasonable than for σ. A delta method (using Taylor's series expansion) may be used to compute inferences on the logarithmic scale and re-transformed to the original scale. The combining rule for $\log \sigma$ uses,

$$U_l = Var(s_l^2)/(4s_l^4)$$

and $e_l = \log s_l$. If (L, U) is the confidence interval for $\log \sigma$ then $(\exp(L), \exp(U))$ is the confidence interval for σ.

For the binary Y, the same strategy used for the mean of the continuous variable can be used for computing the estimate and its standard error. However, better approximations are available for constructing the confidence interval. Two of the most promising approaches are provided here, among many that are reported in the literature. Define,

$$\widehat{\phi}_l = \log \frac{n\bar{y}_l + 1/2}{n(1 - \bar{y}_l) + 1/2}.$$

Then $\widehat{\phi}_l$ is approximately normally distributed with mean $\phi = \log(\mu/(1-\mu))$ and estimated variance $U_l = (n\bar{y}_l + 1/2)^{-1} + (n(1 - \bar{y}_l) + 1/2)^{-1}$. The usual combining rules can be used to obtain, T_{MI}, r_{MI} and ν_{MI} and the confidence interval, $(\widehat{\phi}_L, \widehat{\phi}_U)$ for ϕ. Thus, the confidence interval for μ is $[\exp(\widehat{\phi}_L)/(1 + \exp(\widehat{\phi}_L)), \exp(\widehat{\phi}_U)/(1 + \exp(\widehat{\phi}_U))]$.

The second approach is to use the transformation $\widehat{\theta}_l = \sin^{-1}\sqrt{\bar{y}_l}$ which is approximately normal with mean $\theta = \sin^{-1}\sqrt{\mu}$ and variance $U_l = 0.25/n$. As before, the standard combining rules can be applied to obtain the confidence interval $(\widehat{\theta}_L, \widehat{\theta}_U)$ for θ which leads to $(\sin^2\widehat{\theta}_L, \sin^2\widehat{\theta}_U)$ as the confidence interval for μ.

Many descriptive analyses may also involve comparisons of two or more groups. Such comparisons can be made using the "contrast" feature in *IVE-ware*. For example, suppose that Y is a continuous or binary variable. Also suppose that Z is a binary variable taking the values 0 or 1. Assume the primary analytic interest is in the contrast $\theta = \mu_{[Z=1]} - \mu_{[Z=0]}$. The estimate of this contrast from the completed data l is the difference in the estimates for subgroups formed by $Z = 1$ and $Z = 0$. The combining rule can be applied to this estimate and the standard error and confidence interval can be obtained for θ. This feature can be used to assess any contrast. Suppose Z_1 is gender with $1 = Female$ and $0 = Male$ and Z_2 is race coded as $1 = White$ and $2 = African\ American$. The contrast of interest could be difference of the differences:

$$\left(\mu_{[Z_1=1,Z_2=1]} - \mu_{[Z_1=0,Z_2=1]} \right) - \left(\mu_{[Z_1=1,Z_2=0]} - \mu_{[Z_1=0,Z_2=0]} \right).$$

Though all the contrasts can be estimated through a regression formulation, using the descriptive statistics approach might be more convenient.

2.2 Imputation Task

2.2.1 Imputation of the NHANES 2011-2012 Data Set

The first example uses data from the NHANES 2011-2012 survey. The analysis goal is to perform descriptive analysis of a continuous variable measuring systolic blood pressure and categorical variables indicating high cholesterol and obesity along with a set of linear contrasts by gender, race, and obesity status. Prior to analysis, however, missing data in key variables is addressed by multiple imputation. NHANES uses a complex survey design. The survey design variables are included as predictors and a "design-based" approach is used to correctly estimate variances. Alternative methods are discussed and illustrated in Chapter 9.

Step 1 of the imputation task is to define a list of the variables to be included or excluded from the imputation model. Next, consideration of the variable type (binary, continuous, ordinal, count, or mixed), the amount of missing data and possible bounds or restrictions to be applied in the imputation model is important. The imputation model should include a set of variables that encompass the analysis variables and optimally include more variables than the planned analysis. Usual steps in model building through scatter plots, residual plots etc. should be performed for each variable in the data set.

The following code is typed into the XML editor provided with *IVEware* and then executed by clicking "Run". A number of other methods are possible (see the User's Guide for details) but the XML editor approach is the easiest and is used through out the book. In this example, syntax is presented and explained prior to the analysis results, in a step by step manner.

```
<sas name="Descriptive Analysis Using NHANES Data">
/* Set libname */
libname nhanes 'P:\ive_and_MI_Applications_Book\DataSets
```

```
\nhanes_2011_2012' ;
/* Impute Missing Data */
<impute name="impute_mult1" >
title "Impute NHANES Missing Data" ;
datain  nhanes.nhanes1112_18p_c5 ;
dataout impute_mult1_5 all ;
categorical sdmvstra sdmvpsu edcat age marcat mex othhis
white black other RIAGENDR ;
transfer age18p seqn ;
bounds indfmpir (>=0, <=5)  ;
multiples 5 ;
seed 555 ;
diagnose bpxsy1 ;
run ;
</impute>
/* create indicator of high cholesterol and obese
 from imputed data */
data impute_mult1_5_r ;
set impute_mult1_5 ;
if lbxtc >= 200 then high_cholesterol=1 ; else
high_cholesterol=0 ;
if bmxbmi >= 30 then obese=1 ; else obese=0 ;
run ;
```

The code above uses a "tag" to enclose SAS commands, name the analysis
project, set the "libname" used by SAS, and impute missing data using the
IMPUTE command. The DATAIN commands read the input working data
set and DATAOUT with the ALL option produces a data set containing all
5 multiples from the imputations in a "long" file called "impute_mult1_5".
Another option is to output each imputed data set using the PUTDATA
command (not shown here but demonstrated in the User's Guide).

Other highlights of the IMPUTE syntax are use of a CATEGORICAL
statement for categorical variables, a TRANSFER statement to carry along
2 variables, AGE18P and SEQN but omit from the imputation models, and
use of the default CONTINUOUS option which treats all other variables as
continuous.

Imputation bounds applied are declared in the BOUNDS statement. Here,
use of observed data values of 0 to 5 supply bounds for the imputed values
of the family poverty/income ratio variable. The use of the MULTIPLES
and SEED statements request 5 multiples and a seed value to ensure later
replication of results. Finally, use of DIAGNOSE produces diagnostic plots
for variables listed in the statement.

The imputation section of the program is executed to impute missing data
on the variables INDFMPIR, BPXSY1, BMXBMI, LBXTC, and MARCAT
(family poverty/income ratio, systolic blood pressure measurement #1, Body

Mass Index, total cholesterol, and marital status, respectively). Once imputation is complete, two categorical indicators are created in the SAS data step, OBESE (coded 1 if BMI >=30) and HIGH_CHOLESTEROL (coded 1 if total cholesterol (LBXTC) >=200). Because both of these variables had some missing data, the imputed data is used to create new indicators.

Prior to descriptive analysis of completed data sets, selected output from the imputation process is presented. For example, for just the first of the five imputed data sets, information about observed and imputed observations(note that the double-counted column is mostly for diagnostic purposes when user specified restrictions or bounds are used during the imputation, see the *IVE-ware* users guide for more details) for two variables is presented below. Note that this is just a small sample of all output produced by the software. The output provides a way to visually evaluate the imputation through inspection of the observed versus imputed distributions to identify possible imputation process problems. In addition, use of the DIAGNOSE option for the systolic blood pressure variable (BPXSY1) produces plots to evaluate the imputations (discussed in Section 1.7.1). Based on evaluation of Figure 2.1, a plot of observed v. imputed mean values for systolic blood pressure, no apparent differences are revealed. Given satisfactory imputation results, the focus now turns to descriptive analysis.

Variable INDFMPIR

	Observed	Imputed	Combined
Number	5332	532	5864
Minimum	0	0.0192318	0
Maximum	5	4.95729	5
Mean	2.36687	2.27823	2.35883
Std Dev	1.66749	1.20566	1.63111

Variable BPXSY1

	Observed	Imputed	Combined
Number	5132	732	5864
Minimum	74	67.1915	67.1915
Maximum	238	182.924	238
Mean	123.528	124.089	123.598
Std Dev	18.7225	19.3952	18.807

Figure 2.1: Diagnostics plots to compare the observed and imputed values of systolic blood pressure

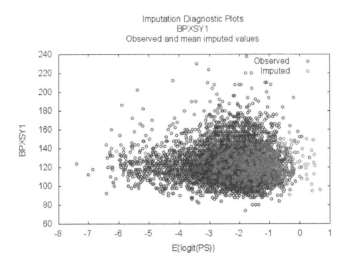

2.3   Descriptive Analysis

2.3.1   Continuous Variable

The goal is to perform a descriptive analysis of the first measurement of systolic blood pressure, among adults age 18 and older, using the 5 completed data sets produced by IMPUTE.

```
<describe name="Mean Systolic Blood Pressure and Proportion
High Cholesterol and Obese" >
title  "Mean Systolic Blood Pressure" ;
datain impute_mult1_5_r ;
stratum sdmvstra ;
cluster sdmvpsu ;
weight  wtmec2yr ;
mean  bpxsy1 ;
table high_cholesterol obese ;
run ;
</describe>

<describe name="Mean Systolic Blood Pressure with Contrast
Gender and Race" >
title  "Systolic Blood Pressure by Gender and Race" ;
```

```
datain impute_mult1_5_r ;
stratum sdmvstra ;
cluster sdmvpsu ;
weight  wtmec2yr ;
mean    bpxsy1 ;
contrast riagendr white ;
run ;
</describe>
```

The previous set of commands first requests an analysis of the continuous variable BPXSY1 (systolic blood pressure) for the total sample as well as by those with high cholesterol and those considered obese. Use of the MEAN, TABLE and STRATUM, CLUSTER, and WEIGHT statements provide a correctly weighted and design-based analysis of the imputed data sets.

The results in Table 2.1 indicate that the estimated mean systolic blood pressure for US adults in 2011-2012 is 121.96 with a standard error of 0.59.

Table 2.1: Descriptive Analysis of Systolic Blood Pressure

Number of Cases Per Multiple	Mean	SE
5,615	121.96	0.59

The second analysis performs linear contrasts of estimated systolic blood pressure by gender and race. The syntax includes STRATUM, CLUSTER, and WEIGHT statements to incorporate the complex sample design features of the NHANES survey, a MEAN statement for analysis of the continuous variable BPXSY1, and a CONTRAST statement for a contrast of mean systolic blood pressure by Gender (coded male, female) and Race/Ethnicity (coded White, Non-White).

Table 2.2: Descriptive Analysis of Systolic Blood Pressure by Gender and Race

Systolic Blood Pressure	Mean	Standard Error	P Value
Male	123.49	0.62	
Female	120.52	0.67	
Contrast	2.97	0.52	0.00*
Non-White	121.46	0.76	
White	122.22	0.77	
Contrast	-0.75	1.02	0.46

Table 2.2 presents the weighted mean, standard error for males (123.49, se=0.62) and females (120.52, se=0.67) along with the linear contrast of 2.97, se=0.52, p=0.00, indicating a significant linear contrast.

The contrast of mean systolic blood pressure for Whites (122.22, se=0.77) versus Non-Whites (121.46, se=0.76), however, is not significant with a difference of (-0.75, se=1.02) and p=0.46.

2.3.2   Binary Variable

The same imputed data sets are now used to perform descriptive analysis of the prevalence of High Cholesterol in US adults as well as differences in prevalence of High Cholesterol by obesity status and gender.

```
<describe name="High Cholesterol with Contrast Gender
 and Obesity Status"  >
title  "High Cholesterol by Obese and Gender with Contrasts" ;
datain impute_mult1_5_r ;
stratum sdmvstra ;
cluster sdmvpsu ;
weight  wtmec2yr ;
table   high_cholesterol ;
contrast riagendr obese ;
run ;
</describe>
</sas>
```

The previous block of command syntax includes use of the TABLE statement rather than the MEAN statement for analysis of a binary variable and also requests a contrast of high cholesterol by gender and obesity status in the CONTRAST statement. The contrast enables a test of the null hypothesis of no association between having a cholesterol measurement of >= 200 and being obese (BMI >=30) and by gender. Finally, the entire program is closed by use of the </sas > statement.

Table 2.3: Prevalence of Obesity and High Cholesterol

Obese	Weighted Proportion	Standard Error
Yes	0.35	0.01
No	0.65	0.01
High Cholesterol		
No	0.58	0.01
Yes	0.42	0.01

The weighted proportions of Table 2.3 suggest that an estimated 35% of US adults are obese while about 42% have high cholesterol (total cholesterol

>= 200).

Table 2.4: Contrasts of High Cholesterol by Gender and Obesity Status

	Proportion (Weighted)	Standard Error	P Value
Female, High Cholesterol	0.46	0.01	
Male, High Cholesterol	0.38	0.01	
Contrast of Gender/High Cholesterol	0.08*	0.02	0.00
Obese, High Cholesterol	0.44	0.02	
Not Obese, High Cholesterol	0.41	0.02	
Contrast of Obese/High Cholesterol	0.03	0.02	0.10

Based on the results from Table 2.4, among women, 46% have high cholesterol while among men, 38% have elevated cholesterol. The contrast is about 8% with a p value of 0.0, indicating a significant association between gender and high cholesterol. A similar analysis shows 44% of those considered obese have high cholesterol while among those that are not obese, 41% have high cholesterol with a contrast of 3% with a non-significant p value of 0.10.

2.4 Practical Considerations

In this chapter we use the XML or SRCShell editor with SAS method of execution. Of course, the XML editor method can also be used with Stata, R, SPSS, and SRCWare (stand-alone *IVEware*). Other methods are covered in the user manual on the *IVEware* website (www.iveware.org). Regardless of the choice of method, the results should be the same; the differences are solely in execution method.

Other considerations are to carefully plan the imputation model and variables to be used to impute missing data before beginning the process. For example, make sure to declare variable type, bounds, and restrictions precisely, and once the imputation is complete, evaluate the quality of the imputations prior to moving to descriptive or regression analyses of complete data sets.

2.5   Additional Reading

Many basic text books discuss construction of descriptive statistics given in Chapter 1. For the use of transformations see Cox and Hinkley (1976). For the analysis of complex survey data see, for example, Graubard and Korn (1999), Lohr(2009), Heeringa et al (2017) and Lumley (2010).

2.6   Exercises

1. Download the Health and Retirement Survey 2012 data set called **EX_HRS_2012**, from the book web site and use multiple imputation to handle missing data. The analysis goal is to estimate mean BMI in the total population and by gender.

   (a) Use a statistical software of your choice to examine the missing data problem. Which variables have missing data and how much missing data is there per variable? What is the overall pattern of missing data?

   (b) Impute any missing data using the IMPUTE command. Make sure to use the observed minimum and maximum as imputation bounds for the BMI variable (R11BMI), use a SEED value to allow replication of results, create M=10 imputations with 5 iterations, and omit ID variables (HHID, PN) from the imputation models.

   (c) Perform a design-based descriptive analysis of mean BMI, using the 10 multiply imputed data sets. Be sure to include the stratification, cluster, and weight variables in the analysis. What is the estimated mean BMI (SE) for the total population?

   (d) Repeat the analysis in part c. but obtain estimated mean BMI and standard error by gender.

   (e) Perform a linear contrast of mean BMI by gender. What is the difference in estimates and standard error for mean BMI for men v. women? Is this difference statistically significant?

2. Download the most recent NHANES demographic and body measurement data from the NHANES web site. Gather the following variables for this exercise: Age, Gender, Race/Ethnicity, Marital Status, Body Mass Index, the MEC 2 year weight, Masked Stratum, Masked PSU, and Total Serum Cholesterol. The goal is to analyze the overall proportion of high cholesterol ($>=200$) among

U.S. adults age 20 and older plus high cholesterol by gender and race/ethnicity. However, missing data must be imputed prior to final analysis.

(a) Use a statistical software of your choice to examine the missing data pattern among those age 20 and older. How many variables have fully observed and missing data? What is the overall pattern of missing data?

(b) Impute any missing data using the IMPUTE command with the number of multiples dependent on your judgement regarding the missing data problem. Use the observed minimum and maximum values as imputation bounds for the continuous variables BMI and total cholesterol, a SEED value to allow replication of results, and omit the case id variable from the imputation models. Make sure to incorporate the complex sample design variables and weight in the imputation.

(c) Produce imputation diagnostics (DIAGNOSE option in IMPUTE) for a few variables with missing data and evaluate the imputation plots. Are there any apparent problems or issues to investigate?

(d) Using the combined imputed data, create an indicator of high cholesterol coded 1 if total cholesterol $>=200$ and 0 otherwise. Create another indicator variable called Black coded 1 if Black and 0 if non-Black. Perform a design-based descriptive analysis using the DESCRIBE command to obtain the proportion of those with high cholesterol in the total sample, and by gender and Black/non-Black. Produce a table of results similar to that from Section 4.3.3. What are your conclusions about estimated high cholesterol in the US population of adults age 20+? What about gender/race contrasts?

3. Exercise 3 also uses the Health and Retirement Survey 2012 data set as in Exercise 1. Prior to beginning this exercise, study Chapter 9 for details on use of the BBDESIGN command for analysis of complex survey data. Use a weighted finite population Bayesian bootstrap approach (implemented in BBDESIGN) to produce an expanded complex sample design data set and then impute missing data within the expanded data set. The goal is to estimate mean BMI in the total sample and compare these results to those from Question 1 using a multiple imputation and design-based approach.

(a) Repeat the examination of the missing data pattern. How many variables have fully observed and missing data? What is the overall pattern of missing data?

(b) Use the BBDESIGN command to expand the data set to represent the complex sample design features such as stratification, clusters, and weights. Create 25 implicates in this step.

(c) Impute any missing data in the expanded data set using the IMPUTE command and request M=5 with 5 iterations. Make sure to use the observed minimum and maximum as imputation bounds for the BMI variable (R11BMI), use a SEED value to allow replication of results, and omit HHID, and PN from the imputation models.

(d) Perform descriptive analysis of mean BMI in the total sample, using the DESCRIBE command but omit the stratum, cluster, and weight statements. Use just the imputed/expanded data set in this step. Make sure to code or calculate the correct variance estimates following the example in the User guide. (Recall that the combining rules for this process are different from MI results). Based on these results, what is the estimated mean BMI (SE) for the population of inference? Why do we omit the design features and weights for the analysis? How do the results compare to those from Question 1?

3

Linear Models

3.1 Introduction

Regression analysis, the backbone of quantitative social and health sciences research, is used to investigate the relationship between a dependent variable Y and a predictor (or a vector of predictors), X. The purpose of this relationship may be to study the effect (or influence) of a particular predictor on explaining the variation in the outcome variable or to develop a system to predict the outcome for subjects yet to be seen from the population.

In a nutshell, a regression model involves specifications of one or more aspects of the conditional distribution of Y given X. A regression function is the conditional mean, $E(Y|X)$, which may be expressed as $g(X;\theta)$ where g is a known function and θ is, possibly, a vector of unknown parameters. In a semi-parametric (non-parametric) formulation, the function g is partially specified (unspecified). This chapter deals with the linear model, a specific form, $g(X;\theta) = X^T\theta$. The linear model refers to the relationship between $E(Y|X)$ and X that is linear with respect to parameters θ. For example, consider a single predictor, Z and $g(X;\theta) = \theta_o + \theta_1 Z + \theta_2 Z^2 = X^T\theta$ where $X = [1 \ Z \ Z^2]^T$ is a vector of predictors and $\theta = [\theta_o \ \theta_1 \ \theta_2]^T$ is a vector of regression coefficients. This is a linear model but a nonlinear relationship between $E(Y|X)$ (more specifically $E(Y|Z)$) and Z.

The second aspect of the conditional distribution is the variance,

$$Var(Y|X,\sigma) = h(X;\sigma),$$

where h is a known function of predictors and unknown parameters, σ. For many distributions, the mean and variance are related to each other, hence h will be a function of g. As in the case of g, h may be partially specified or unspecified. In this chapter, we consider the case $h(X;\sigma) = \sigma^2$ or $\sigma^2 h(X)$ where $h(X)$ is a known function of X.

Typically, the mean and variance functions are sufficient in a regression analysis. In some instances, the full conditional distribution may be specified. For example, $f(y|x)$ is a normal density function with the above specified mean and variance function. In general, a regression model may be specified as

$$Y = g(X;\theta) + \epsilon\sqrt{h(X;\sigma)}$$

where ϵ is a random variable having a distribution with density function f, mean 0 and variance 1. For example, a normal linear model with a constant variance is

$$Y = X^T\theta + \epsilon\sigma$$

where $\epsilon \sim N(0,1)$ whereas the same model with a non-constant variance is

$$Y = X^T\theta + \epsilon\sigma\sqrt{h(X)}.$$

3.2 Complete Data Inference

3.2.1 Repeated Sampling

A popular approach for the estimation of the unknown parameter θ (for a given σ) based on a sample, $\{(y_i, x_i), i = 1, 2, \ldots, n\}$ of size n is using the method of least squares. The goal is to find θ that minimizes the objective function,

$$\sum_i^n \left(\frac{(y_i - g(x_i;\theta))^2}{h(x_i;\sigma)} \right).$$

Another set of least squares equations are usually developed to estimate σ but will not be necessary for the type of models used in this chapter.

For the linear model with the constant variance, the method of least squares is equivalent to minimizing the familiar, residual sum of squares,

$$\sum_i (y_i - x_i^T\theta)^2$$

which leads to $\widehat{\theta} = (X^TX)^{-1}X^TY$ where X is a $n \times p$ matrix with rows formed by stacking the row vectors $x_i, i = 1, 2, \ldots, n$ one below the other and Y is the corresponding $n \times 1$ vector.

The residual sum of squares is $\sum_i (y_i - x_i^T\widehat{\theta})^2$ which has $n - p - 1$ degrees of freedom where p is the number of predictors (not including the intercept). It can be shown that

$$\widehat{\sigma}^2 = \frac{\sum_i (y_i - x_i^T\widehat{\theta})^2}{n - p - 1}$$

is an unbiased estimate of σ^2.

The covariance matrix of the regression coefficient, $\widehat{\theta}$, is $\widehat{\sigma}^2(X^TX)^{-1}$. The sampling variance of $\widehat{\sigma}^2$ is $a/(n-p-1) - \widehat{\sigma}^4(n-p-4)/((n-p-1)(n-p-2))$ where $a = \sum_i e_i^4/(n-p-1)$ and $e_i = y_i - x_i^T\widehat{\theta}$ is the residual corresponding to the observation $i = 1, 2, \ldots, n$.

If the conditional distribution of Y given X is normal with mean $X^T\theta$ and variance σ^2, then it can be shown that the sampling distribution of $\widehat{\theta}$ given θ, σ

and X is multivariate normal with mean θ and covariance matrix $\sigma^2(X^TX)^{-1}$. The sampling distribution of $\widehat{\sigma}^2$ given σ^2 and X is $\sigma^2\chi^2_{n-p-1}/(n-p-1)$, a scaled chi-square distribution with $n-p-1$ degrees of freedom. From these two sampling distributions, it follows that the sampling distribution of $\widehat{\sigma}^{-1}C(\widehat{\theta}-\theta)|\theta, X$ is a multivariate t with $n-p-1$ degrees of freedom where C is the Cholesky decomposition such that $CC^T = (X^TX)^{-1}$. This result forms the basis for constructing confidence intervals for the components of θ or testing hypothesis concerning values of θ.

For a nonconstant variance of the type $Var(Y|X) = h(X)\sigma^2$, the least squares approach is equivalent to minimizing,

$$\sum_i^n \frac{(y_i - x_i^T\theta)^2}{h(x_i)} = \sum_i^n \left[\frac{y_i}{\sqrt{h(x_i)}} - \frac{x_i^T}{\sqrt{h(x_i)}}\theta\right]^2 = \sum_i (y_{i*} - x_{i*}^T\theta)^2,$$

where $y_{i*} = y_i/\sqrt{h(x_i)}$ and $x_{i*} = x_i/\sqrt{h(x_i)}$. That is, a regression analysis with non-constant variance can be re-expressed as constant variance regression analysis with an appropriate scaling.

It is often difficult to model the variance, $Var(Y|X)$. One approach for drawing inferences about the regression function $E(Y|X,\theta)$, is to use a repeated replication technique for computing the sampling variance of any estimate, $\widehat{\theta}$, of θ. In *IVEware*, the Jackknife Repeated Replication (JRR) technique or method can be used to estimate the sampling variance, designed for complex surveys, by treating each individual as a "cluster".

Let $\widehat{\theta}_{(-i)}$ be the estimate without the subject $i = 1, 2, \ldots, n$. A pseudo value is defined as

$$\widehat{\theta}_i^* = n\widehat{\theta} - (n-1)\widehat{\theta}_{(-i)}$$

The bias corrected Jackknife estimate is defined as

$$\widehat{\theta}_{JK} = \sum_i \widehat{\theta}_i^*/n$$

with its sampling variance estimated as

$$v_{JK} = \frac{1}{n(n-1)}\sum_i (\widehat{\theta}_i^* - \widehat{\theta}_{JK})(\widehat{\theta}_i^* - \widehat{\theta}_{JK})^T$$

$$= \frac{n}{n-1}\sum_i (\widehat{\theta}_{(-i)} - \widehat{\theta})(\widehat{\theta}_{(-i)} - \widehat{\theta})^T.$$

The Jackknife approach can be quite computationally intensive for a large sample size. One can reduce the computational burden by creating random non-overlapping k subsets of n observations and using these random subsets as "clusters". This feature will be useful in longitudinal analysis where one can model just the mean function and treat each individual as a cluster to obtain valid variance estimates without specifying the correct variance function.

3.2.2 Bayesian Analysis

For a Bayesian analysis (normal model with constant variance), a prior distribution for (θ, σ) is needed. In many practical situations, this prior distribution is typically diffuse relative to likelihood (constructed based on the conditional distribution of Y given X, θ, σ). It is convenient to express this prior through a so-called non-informative improper prior, Jeffreys prior, $Pr(\theta, \sigma) \propto \sigma^{-1}$. Under this prior, it can be shown that,

1. The posterior distribution of θ given $\widehat{\theta}, \sigma$ and X is normal with mean $\widehat{\theta}$ and the covariance matrix $\sigma^2(X^TX)^{-1}$.

2. The posterior distribution of σ^2 given $\widehat{\sigma}^2$ is given by

$$\frac{(n-p-1)\widehat{\sigma}^2}{\sigma^2}|\widehat{\sigma} \sim \chi^2_{n-p-1},$$

or

$$\sigma^2|\widehat{\sigma} \sim \frac{(n-p-1)\widehat{\sigma}^2}{\chi^2_{n-p-1}}.$$

3. From (1) and (2), it follows that the posterior distribution of θ given $\widehat{\theta}$ and X (that is, integrating out σ) is a multivariate t distribution with location $\widehat{\theta}$ and the scale matrix $\widehat{\sigma}^2(X^TX)^{-1}$ and degrees of freedom $n-p-1$.

The posterior probability statements about θ can be constructed using the above posterior distribution. Let θ_j and $\widehat{\theta}_j$ be the jth component of θ and $\widehat{\theta}$, respectively and u_j be the jth diagonal element of the covariance matrix $U = \widehat{\sigma}^2(X^TX)^{-1}$. The $100(1-\alpha)\%$ highest posterior density interval (credible interval) is given by $\widehat{\theta}_j \pm t_{n-p-1,\alpha/2}\sqrt{u_j}$ where $t_{n-p-1,\alpha/2}$ is the quantile of the t distribution with $n-p-1$ degrees of freedom corresponding to the probability $\alpha/2$. Numerically, this is exactly the same as the $100(1-\alpha)\%$ confidence interval from the repeated sampling perspective.

3.3 Comparing Blocks of Variables

The regression coefficient, say θ_1 in a regression model with two variables, $g(X, \theta) = \theta_o + \theta_1 X_1 + \theta_2 X_2$ measures an average expected difference in the outcome, Y, for a one unit positive difference in X_1 when X_2 is held constant. Thus, each regression coefficient is measuring the "unique" impact of the corresponding variable with other variables being equal or held constant. Note that, this interpretation is not valid when predictors include non-linear terms or interaction terms. For example, suppose that $g(X, \theta) = \theta_o + \theta_1 X_1 + \theta_2 X_2 + \theta_3 X_1 \times X_2$; then it is not possible to change X_1 without automatically changing

$X_3 = X_1 \times X_2$. If θ_3 is large then it is not possible to describe the relationship between Y and X_1 (or X_2) using a single line.

Sometimes, it may be of interest to assess the impact of a block of variables $Z_1 = (X_1, X_2, \ldots, X_q)$ where $Z_2 = (X_{q+1}, \ldots, X_p)$ is the remainder of the variables in the "Full" model. The "reduced" model has just Z_2 as the predictors, that is, the regression coefficients corresponding to Z_1 are set to zero. Let RSS_F be the residual sum of squares for the full model and RSS_R be the residual sum of squares for the reduced model. The contribution of Z_1 is measured by the reduction in residual sum of squares $SS(Z_1) = RSS_R - RSS_F$ which is attained by fitting q parameters. That is, on the average, the q variables contribute, $s_N = SS(Z_1)/q$, towards the reduction in the residual sum of squares. As a contrast, the unexplained residual variance is $s_D = RSS_F/(n - p - 1)$ (including the intercept term). Heuristically, the block of variables Z_1 contribute towards our understanding of the variation in Y, if s_N is considerably larger than s_D. The ratio, $F = s_N/s_D$, is a measure of the "worth" of Z_1. Under the hypothesis that Z_1 is irrelevant, the sampling distribution of F has an F-distribution with q and $n - p - 1$ degrees of freedom. Thus, if the computed value of F is large (or in the tail of the distribution of F with $\nu_N = q$ and $\nu_D = n - p - 1$ degrees of freedom) then it is worth including Z_1 in the model.

Sometimes, it is of interest to assess the overall fit of the model. Two popular measures are R^2 and adjusted R^2. Suppose that the reduced model has just the intercept term and the full model has both (Z_1, Z_2) and let F be the corresponding F-statistic. The quantity

$$R^2 = \frac{\nu_N F}{\nu_N F + \nu_D}$$

is the familiar, proportion of the total variation in Y explained by (Z_1, Z_2) and the adjusted R^2 is

$$R^2_{adj} = 1 - (1 - R^2)(n - 1)/(n - p - 1).$$

3.4 Model Diagnostics

Developing good regression models is an iterative process. Start with a working model, use the residuals to assess the model fit, refine the models, assess the fit again until a satisfactory model for both the mean and the variance functions are obtained. This section describes some key diagnostic plots useful for model building purposes.

The model diagnostics can be be classified into the following four categories:

1. *Assess the fit of the mean function.* Scatter plots of the outcome

versus predictors is the first step towards developing the model.
The second step is to plot residuals (from the current model) versus
predictors to detect non-linearity and potential interaction terms.
For a good fit, the residual scatter plots should resemble a random
scatter around the horizontal line at zero value of the residual.

2. *Assess the Constancy of Variance.* A scatter plot of the residuals
   (from the current model) versus the predicted values is useful to
   assess heteroscedasticity. All the scatter plots listed under (1) can
   also be useful to assess the constancy of variance of the residuals
   with respect to individual predictors.

3. *Normality of Residuals.* Histograms and normal quantile plots of
   the residuals are useful to assess whether it is reasonable to assume
   a normal distribution for the residuals. Generally, the least squares
   approach is fairly robust to modest departures from normality. The
   goal is to detect large departures from a bell-shaped distribution.

4. *Influence Diagnostics.* Cooks distance,

$$D_i = (\widehat{\beta}_{-i} - \widehat{\beta})^T \widehat{V}^{-1} (\widehat{\beta}_{-i} - \widehat{\beta}),$$

where $\widehat{\beta}_{-i}$ is an estimate of β without the observation i and $\widehat{V}$ is the
covariance matrix, is useful to assess the influence of observation i.
Another measure is leverage h_i which is the diagonal element of
the "Hat" matrix, $H = X(X^T X)^{-1} X^T$. Eigenvalues of the matrix
$X^T X$ are useful to detect collinearity. Eigenvalues close to zero are
indicative of linear dependencies among the predictors.

3.5   Multiple Imputation Analysis

This section explains how to perform a regression analysis when some covariates and predictors have missing values. First, use the sequential regression
approach to impute the missing values using as many predictors as possible
even though the ultimate analysis may involve only a subset of variables. For
more details about the imputation issues see Raghunathan (2016).

3.5.1   Combining Point Estimates

Suppose that $\widehat{\theta}_l$ is the estimated regression coefficient from imputed data
$l = 1, 2, \ldots, M$ and let U_l be the corresponding covariance estimate. The
multiple imputation estimate of θ is the average, $\widehat{\theta}_{MI} = \sum_l \widehat{\theta}_l / M$ and the
covariance matrix,

$$T_{MI} = \sum_l U_l / M + (1 + M^{-1}) \sum_l (\widehat{\theta}_l - \widehat{\theta}_{MI})(\widehat{\theta} - \widehat{\theta}_{MI})^t / (M - 1).$$

$$= \bar{U}_{MI} + B_{MI}$$

The square root of the diagonal element of T_{MI} are the multiple imputation standard errors.

Note that the matrix T_{MI} may not be positive definite if M is smaller than p, the dimension of θ. However, the following approximation may be useful. If the effect of missing data is similar for all parameters, then the Eigenvalues of B_{MI} relative to T_{MI}, the measures of fraction of information for θ, will be approximately the same, leading to an approximation,

$$T_{MI} = (1 + g_{MI})\bar{U}_{MI}$$

where $g_{MI} = (1 + M^{-1})Tr(B_{MI}\bar{U}_{MI})^{-1}/p$. If this approximation is not reasonable then one may have to choose M large enough to reliably estimate the full covariance matrix T_{MI}, an essential quantity for multivariate hypothesis testing.

3.5.2 Residual Variance

Under normality, the completed data estimate of the residual variance, $\hat{\sigma}_l^2$, is independently distributed (conditional on the observed data), with the following distribution:

$$\frac{(n - p - 1)\hat{\sigma}_l^2}{\sigma^2} \sim \chi^2_{n-p-1}$$

or equivalently,

$$\sigma^{-2}|D_l \sim \frac{\chi^2_{n-p-1}}{(n - p - 1)\hat{\sigma}_l^2}$$

The completed data posterior mean and variance of σ^{-2} are $\hat{\sigma}_l^{-2}$ and $U_l = 2(n - p - 1)^{-1}\hat{\sigma}_l^{-4}$. Applying the combining rules, the multiple imputation estimate of σ^{-2} is $\hat{\sigma}_{MI}^{-2} = \sum_l \hat{\sigma}_l^{-2}/M$ and its posterior variance is

$$T_{MI} = \sum_l U_l/M + (1 + M^{-1})\sum_l (\hat{\sigma}_l^{-2} - \hat{\sigma}_{MI}^{-2})^2/(M - 1).$$

The normal approximation for the terms involving square may not be reasonable. An alternative approach is suggested for inferring about the variance components. Note that, the exact posterior distribution of σ^{-2} is that of a linear combination of independent chi-square random variables. Using the approximation discussed in Satterthwaite (1943), this posterior distribution can be expressed as

$$\sigma^{-2}|D_{obs} \approx a\chi_b^2$$

where $ab = \hat{\sigma}_{MI}^{-2}$ and $2a^2b = T_{MI}$. Equivalently, $a = 0.5 \times T_{MI} \times \hat{\sigma}_{MI}^{-2}$ and $b = 2(\hat{\sigma}_{MI}^{-2})^2/T_{MI}$.

An alternative approach is to use a normal approximation for the completed data posterior distribution of $\log \sigma$ with mean $\log \hat{\sigma}_l$ and variance

$U_l/(4\widehat{\sigma}_l^4)$ which leads to

$$\log \sigma | D_{obs} \approx N(\log \widehat{\sigma}_{MI}, T_{MI})$$

where $\log \widehat{\sigma}_{MI} = \sum_l \log \widehat{\sigma}_l / M$ and variance,

$$T_{MI} = (\sum_l U_l/\widehat{\sigma}_l^4/M)/4 + (1 + M^{-1})\sum_l (\log \widehat{\sigma}_l - \log \widehat{\sigma}_{MI})^2/(M-1)$$

The confidence/credible intervals for σ^{-2} or $\log \sigma$ can be obtained using the approximate scaled chi-square or normal approximation, respectively. The end points can be re-transformed to the original scale (σ^2).

3.6   Example

The linear regression example demonstrates the process of multiple imputation and estimation (with diagnostic plots and multivariate tests) of linear regression. The model includes the continuous outcome total cholesterol measurement predicted by continuous BMI, age in years, gender, and race/ethnicity, using NHANES 2011-2012 data for US adults age 18+. *IVEware* with SAS is used in this example. As in Chapter 2, the analysis is done using a design-based approach though use of the weighted Bayesian Bootstrap method is another excellent option, see Chapter 9 for details and examples.

Imputation is done with IMPUTE, the output data sets are then analyzed with REGRESS (including optional diagnostic plots) and linear regression while accounting for increased variability due to multiple imputation and complex sample design variance estimation. Post estimation, use of SAS PROC SURVEYREG and PROC MIANALYZE are used to produce multivariate tests of specific variables' contribution to the overall model. Finally, calculation of R Squared and Adjusted R Squared statistics is done in the SAS data step using F Test information from PROC MIANALYZE.

3.6.1   Imputation

```
<sas name="Linear Regression Using NHANES Data">
/* Set libname */
libname nhanes 'P:\ive_and_MI_Applications_Book
\DataSets\nhanes_2011_2012' ;

/* examine missing data */
proc means data=nhanes.nhanes1112_adults_26may2016
 n nmiss mean ;
```

```
run ;

/* Impute Missing Data */
<impute name="impute_mult1" >
title "Impute NHANES Missing Data Adults 18Plus" ;
datain   nhanes.nhanes1112_adults_26may2016 ;
dataout impute_mult1 ;
categorical sdmvstra sdmvpsu edcat age marcat ridreth1
 RIAGENDR ;
transfer seqn ;
bounds indfmpir (>=0, <=5)   ;
multiples 5;
iterations 10 ;
seed 555 ;
run ;
</impute>

/* extract the remaining four multiply imputed datasets */
<putdata name="impute_mult1" mult="2" dataout="impute_mult2" />
<putdata name="impute_mult1" mult="3" dataout="impute_mult3" />
<putdata name="impute_mult1" mult="4" dataout="impute_mult4" />
<putdata name="impute_mult1" mult="5" dataout="impute_mult5" />
```

The initial set of commands (above) evaluate the missing data problem using
SAS PROC MEANS, impute missing data with the IMPUTE command, and
output 5 completed data sets using the PUTDATA command. The imputation includes the BOUNDS statement for the INDFMPIR variable, a SEED
statement, TRANSFER and an ITERATIONS statement to specify imputation model details. The PUTDATA commands extracts 5 SAS data sets of
completed data for subsequent analysis.

3.6.2   Parameter Estimation

```
/* Linear Regression Using Imputed Data Sets */
<regress name="Linear Regression Total Cholesterol Regressed on
 BMI Gender Age and Race" >
title  "Linear Regression Total Cholesterol Predicted by BMI
Gender Age and Race" ;
datain impute_mult1 impute_mult2 impute_mult3 impute_mult4
 impute_mult5 ;
stratum sdmvstra ;
cluster sdmvpsu ;
weight  wtmec2yr ;
link linear ;
```

```
dependent lbxtc ;
categorical riagendr ridreth1 ;
predictor bmxbmi age riagendr ridreth1 ;
estimates
black_male_age30_bmi25: Intercpt (1) bmxbmi(25) age (30)
riagendr (1) ridreth1 (4) /
black_female_age30_bmi25: Intercpt (1) bmxbmi(25) age (30)
riagendr (2) ridreth1(4) ;
plots   outplots ;
run ;
</regress>
```

In the second block of code, parameter estimation is performed by linear re-gression with the REGRESS command with the 5 completed data sets output from IMPUTE. Incorporation of the STRATUM, CLUSTER, and WEIGHT statements accounts for the complex sample design and weighting features of the NHANES survey. The dependent variable, LBXTC, is declared on the DEPENDENT statement, model predictors on the PREDICTOR statement, predicted values for certain types of respondents on the ESTIMATE state-ment, and diagnostic plots are generated by the PLOTS statement (see Figure 3.1 below).

Figure 3.1: Regression diagnostics plots

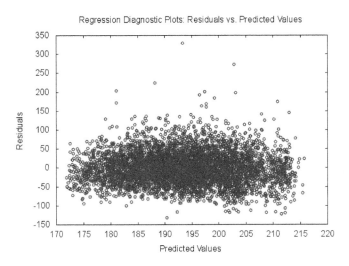

Based on Figure 3.1, a diagnostic plot of Predicted Values v. Residuals, we have little concern about the overall regression model fit. This is one of a number of plots produced by the PLOT option in the REGRESS command.

Table 3.1 presents results from the linear regression including parameter es-timates, standard errors, and CI's for BMI, Age, Gender, and Race/Ethnicity.

Table 3.1: Estimates of Regression Parameters for Total Cholesterol

Variable	Estimate	Standard Error	Confidence Interval
Intercept	170.766	2.721	(164.945, 176.586)
BMXBMI	0.217	0.072	(0.062, 0.374)
Age	0.384	0.056	(0.264, 0.503)
Male	-8.702	1.247	(-11.370, -6.035)
Mexican	3.931	2.845	(-2.155, 10.017)
Other Hispanic	4.426	3.388	(-2.821 , 11.674)
White	5.128	2.304	(0.198, 10.057)
Black	-2.247	2.657	(-7.930,3.436)

The results suggest that all but Race/Ethnicity are significant predictors of estimated Total Cholesterol in the U.S. adult population in 2011-2012.

Table 3.2: Estimates of Predicted Total Cholesterol for Certain Types of Individuals

Characteristics	Estimate	Standard Error
Black, Male, Age 30, BMI=25	194.748	9.616
Black, Female, Age 30, BMI=25	186.046	9.699

Table 3.2 presents predicted total cholesterol levels for Black males and females age 30 with a BMI of 25. The estimates are higher for males as compared to females with the same characteristics. These results are produced by the ESTIMATE statement in the previous REGRESS command of *IVEware*.

3.6.3 Multivariate Hypothesis Testing

A multivariate test of zero contribution to the linear model is demonstrated by use a test of two predictors, indicators of being Male and White.

```
/* Sort by _mult_ before running SURVEYREG */
data impute_mult1_5_r ;
set impute_mult1 impute_mult2 impute_mult3 impute_mult4
 impute_mult5  ;
/* create a new variable for PROC SURVEYREG AND PROC MIANALYZE */
_imputation_= _mult_ ;
if riagendr=1 then male=1 ; else male=0 ;
run ;
proc sort ;
 by _imputation_ ;
run ;

/* SAS PROC SURVEYREG with PROC MIANALYZE to do F Test and
```

```
Multivariate Tests*/
title "SAS SURVEYREG with PROC MIANALYZE for Multivariate Tests";
proc surveyreg data=impute_mult1_5_r ;
strata sdmvstra ;
cluster sdmvpsu ;
weight wtmec2yr ;
class ridreth1 ;
by _imputation_ ;
model lbxtc = bmxbmi age male ridreth1 / solution covb ;
ods output parameterestimates=outparms  covb=outcovb ;
run ;
proc print data=outparms ;
run ;

title "Compressed Parameter" ;
* modify name of race/eth to match outcovb ;
data outparms1 ;
 set outparms ;
 parameter=compress(parameter) ;
run ;
proc print ;
run ;

data outcovb1;
 set outcovb ;
 parameter=compress(parameter) ;
 ridreth11=ridreth1_1; ridreth12=ridreth1_2 ;
 ridreth13=ridreth1_3 ; ridreth14=ridreth1_4 ;
 ridreth15=ridreth1_5 ;
run ;
title "Compressed Covb Parameter" ;
proc print ;
run ;

/*use OUTPARMS and OUTCOVB in PROC MIANALYZE for Multivariate
Test of Race */
proc mianalyze parms=outparms1 covb=outcovb1 ;
  modeleffects intercept bmxbmi age male ridreth11 ridreth12
  ridreth13 ridreth14 ;
  test bmxbmi,age, male ,ridreth11, ridreth12, ridreth13,
   ridreth14 / mult ;
  test male, ridreth14 / mult;
run ;
```

The above commands use the concatenated imputed data set output from
IMPUTE, execute linear regression for each of 5 multiples by SAS PROC

SURVEYREG and save parameter estimates and standard errors in an output data set for use in PROC MIANALYZE. Note the use of the SAS Data Step to remove white space in the output parameter estimates stored in the variable called "parameter". This is accomplished with the COMPRESS function of SAS and is needed prior to use of PROC MIANALYZE. The MIANALYZE procedure offers MI and design-adjusted multivariate tests for specified variables. The code includes a TEST statement with the option / MULT to perform a joint test of the specified variables' contribution to the model.

Table 3.3: Multivariate Inference for Male and Black Race for Linear Regression Model Predicting Total Cholesterol

Average Relative Increase in Variance	Num DF	Den DF	F	Pr>F
0.069	2	570.70	30.95	<.0001

3.6.4 Combining F-statistics

An F-test of the hypothesis of zero contribution to the model for all model predictors is performed using PROC SURVEYREG and PROC MIANALYZE (see above syntax). This step uses a similar TEST statement as presented for a "reduced" model but in this case, all predictor variables are included in the TEST statement.

Table 3.4: F-Test for Linear Regression Model Predicting Total Cholesterol

Average Relative Increase in Variance	Num DF	Den DF	F	Pr>F
0.121	7	1812.30	34.48	<.0001

Results from Tables 3.3 and 3.4 suggest that indicators of being Male and Black significantly contribute to the linear model and the overall F test indicates that all 7 predictors are also significantly different than zero contribution to the model.

3.6.5 Computation of R^2 and Adjusted R^2

The R^2 and Adjusted R^2 statistics can be calculated following the formulae presented previously in this chapter or on page 113 of Raghunathan (2016). In this example, we use the F test information from Section 3.6.3 and calculate the R^2 and Adjusted R^2 statistics in the SAS data step.

```
/* R Squared and Adjusted R Squared using information from
PROC MIANALYZE */
```

```
data rsq ;
 * rsq = dfnum*F/(dfnum*F + dfdenom) ;
 rsq =7*34.48 / (7*34.48 + 1812.30) ;
 * adjusted R sq = 1-(1-Rsq)*(n-1)/ (n-p-1) ;
 adjrsq = 1-(1-rsq)*(5864-1) /(5864-7-1);
run ;
proc print ;
 var rsq adjrsq ;
run ;
</sas>
```

The data step syntax performs the needed calculations and finally, closes the session by use of the $</sas>$ tag.

Table 3.5: R^2 and Adjusted R^2 for Linear Regression Model Predicting Total Cholesterol

R^2	Adjusted R^2
0.118	0.116

The overall R^2 and the Adjusted R^2 are both about .12, indicating about 12% of the variability in the outcome is explained by the seven independent variables in this model.

3.7   Additional Reading

For a practice oriented book on regression analysis, see Gelman and Hill (2006). There are other classic books providing a comprehensive review of regression analysis such as Draper and Smith (1998), Weisberg (2013), Neter, Kutner, Nachtsheim, and Wasserman (1996). Some of the early work on missing values in regression analysis is described in Afifi and Elashoff (1967) and for the missing values in ANOVA, Allan and Wishart (1930). See also Dodge (1985). Rubin (1976b, 1976c) discusses missing values in the outcome and predictors. A comprehensive review of regression analysis with missing predictors is given in Little (1992). More recent references include Von Hippel (2007).

3.8 Exercises

1. Consider the following data regarding physical fitness, collected from men in a fitness course at N.C. State University (data obtained from the SAS Institute example data sets).

   (a) Read the data below into a data set ready to be imputed and analyzed by *IVEware*, using your chosen statistical software. For example, the SAS code below illustrates how to read raw data into a SAS data set:

   ```
   /* Data on Physical Fitness:
   These measurements were made on men involved
   in a physical fitness course at N.C. State University.
   Certain values have been set to missing and the resulting
   data set has an arbitrary missing pattern. Only selected
   variables of Oxygen (intake rate, ml per kg body weight
   per minute), Runtime (time to run 1.5 miles in minutes),
   RunPulse (heart rate while running) are used. */

   data Fitness1;
   input Id Oxygen RunTime RunPulse @@;
   datalines;
   1 44.609  11.37  178     2 45.313  10.07  185
   3 54.297   8.65  156     4 59.571    .      .
   5 49.874   9.22    .     6 44.811  11.63  176
   7  .      11.95  176     8   .     10.85    .
   9 39.442  13.08  174    10 60.055   8.63  170
   11 50.541   .      .    12 37.388  14.03  186
   13 44.754  11.12  176   14 47.273    .      .
   15 51.855  10.33  166   16 49.156   8.95  180
   17 40.836  10.95  168   18 46.672  10.00    .
   19 46.774  10.25    .   20 50.388  10.08  168
   21 39.407  12.63  174   22 46.080  11.17  156
   23 45.441   9.63  164   24   .      8.92    .
   25 45.118  11.08    .   26 39.203  12.88  168
   27 45.790  10.47  186   28 50.545   9.93  148
   29 48.673   9.40  186   30 47.920  11.50  170
   31 47.467  10.50  170
   ;
   run;
   ```

   (b) The analytic goal is to use linear regression to predict time to run 1.5 miles by oxygen intake and running heart rate. Begin by performing an analysis of the extent of the missing data problem and the overall missing data pattern. Use your software of choice for this step. How would you describe the missing data pattern?

(c) What percentage of the data would be lost if performing a complete case analysis? How much missing data exists for each variable and what are the variable types in this data set?

(d) Impute the missing data using 4 different M= options. Begin with M=5 and then increase to 20, 25, and the percentage of cases with any missing data. Save each of the output imputed data sets in a "long" format for use in linear regression analyses to come.

Use the following options in the imputation: 1. transfer the ID variable during the imputation process, 2. use a seed value to ensure replication of results, 3. use built-in imputation diagnostics to evaluate the imputations, and 4. use the ALL option on the output data sets of imputed data.

(e) Prepare scatter plots prior to linear regression to ensure that linear regression is a viable model choice. Also, check the distribution of the dependent variable (RUNTIME) for violation of the normality assumption. Is linear regression a good choice for these variables? Do any variables require transformations?

(f) Use each of the imputed data sets from d. plus the original data (non-imputed) and perform linear regression where Running Time is regressed on Oxygen Intake and Run Pulse. Prepare a table including Parameter Estimates, SEs, T Tests, and p values. Do the results change as the number of imputed data sets increases? How do the imputed results compare to the Complete Case Analysis? Would your conclusions change if you impute missing data?

2. Continue with the data set used in Chapter 2, Exercise 2 (based on the most recent NHANES data with selected variables). The goal is to perform linear regression of total cholesterol regressed on gender and Body Mass Index and to compute a number of model fit statistics such as R Squared, Adjusted R Squared, and F test statistics. Given the complex sample and weights included in the NHANES data, make sure to incorporate these features in the imputation models and subsequent analyses.

(a) Repeat the examination of the extent and pattern of missing data. How much missing data exists and what is the pattern of missing data?

(b) Impute any missing data in the data set using M=(a number of your choice). Why did you choose this number of multiples? Why include the dependent variable in the imputation model? Make sure to consider possible restrictions or bounds needed in your imputation. Did you use either of these statements and if so, why?

(c) Using your imputed data sets, prepare a table of regression results from the linear model of total cholesterol regressed on gender and BMI. Make sure to include Parameter Estimates, SE's, T Test, and P values plus an overall F Test, R Squared and Adjusted R Squared (using the correct MI combining rules presented previously in this chapter).

3. This exercise uses selected variables from the NHANES 2005-2006 data set. The variables are drawn from the interview portion of the survey (demographic, design variables and weights) and the Medical Examination Component (blood pressure). The goal is imputation of missing data followed by linear regression to predict diastolic blood pressure by age and gender.

(a) Download the data set called **EX_SUBSET_NHANES_0506** from the book web site and if needed, convert to a data set appropriate for use in your software of choice. Note, this data set is restricted to those age 18+ and contains a subset of variables for imputation and analysis, n=5,563. Age has been centered by subtracting mean age in the subpopulation of adults (45.60) from the original age variable.

(b) Examine the extent and pattern of missing data. Impute missing data using M=(number of your choice), a seed value, bounds for the blood pressure variable (your choice), transfer the case ID and age 18+ indicator, and request imputation diagnostics. Do the diagnostics suggest any problems with the imputation and if so, what might you do to address the problems?

(c) Using the imputed data sets from b., run a "preliminary" regression model of Diastolic Blood Pressure regressed on gender and centered age. Examine the residual*centered age plot and evaluate the results. What does the plot suggest?

(d) Add a squared age term to the model and repeat step c. Do you see improvement in the residual*centered age plot when the squared term is added?

(e) Run your "final" linear regression and present design-based and MI combined Parameter Estimates, SE, T Tests, and p Values. Provide a short paragraph, as for publication, interpreting these results including describing the imputation process used and how *IVEware* deals with both MI variability and complex sample design features in the analysis.

4

Generalized Linear Model

4.1 Introduction

The linear models discussed in the previous chapter, mostly assuming normally distributed residuals, can be generalized to handle non-normal outcome variables. The density functions of an exponential family of distributions is of the form,

$$f(y|\beta) = h(y) \exp(g(\beta)^T S(y) - A(\beta))$$

where β is a vector of parameters, $h(), g(), S()$ and $A()$ are known functions. Many distributions such as normal, binomial, Poisson, gamma, etc. are members of this family. For more details about this family of distributions, see McCullagh and Nelder (1989).

A generalized linear model uses this family of distributions, for the conditional distribution of Y given X, $pr(Y|X)$, by specifying the mean and variance functions as follows. Let $\mu(x) = E(Y|X, \theta)$ be the mean and $V(x) = Var(Y|X, \theta, \sigma)$ be the variance function. Assume that $f(\mu(x)) = x^T \theta$ and $V(x) = \sigma^2 V(\mu)$. The function f is called the "link" function. For a binomial distribution, with f as the logit function, $\sigma = 1$ and $V(\mu) = \mu(1 - \mu)$ results in the familiar logistic regression model. Similarly, f as the logarithm, $\sigma = 1$ and $V(\mu) = \mu$ results in a Poisson regression model. There are many choices available to handle several types of outcomes and distributions.

The method of maximum likelihood is typically used to estimate the unknown parameters θ (and σ). These reduce to solving $p + 1$ equations (corresponding to p predictors and the intercept term) of the form,

$$\sum_i \left(\left(\frac{\partial \mu(x_i)}{\partial \theta} \right)^T \left(\frac{y_i - \mu(x_i)}{V(\mu(x_i))} \right) \right) = 0$$

An iterative approach is used to solve the equations employing the method of iteratively re-weighted least squares.

The modeling approach has been generalized to handle deviations from restrictions imposed by the distributional assumptions. For example, the Poisson distribution assumes that the mean and variances are exactly equal. Obviously, the data may indicate under-dispersion or over-dispersion. These models can be fit by assuming that σ is not equal to 1 but estimated from the data. For this case, there is no corresponding specific distribution, and the approach is called the quasi-likelihood method for fitting generalized regression models.

Let $\widehat{\theta}$ be the estimate of the regression parameters. The covariance matrix of $\widehat{\theta}$ is of the form $\widehat{V} = \widehat{\sigma}^2(X^T W X)^{-1}$ where W is a diagonal matrix and is a function of $\widehat{\theta}$ and X. For certain members of the exponential family, the covariance matrix is also equivalent to

$$\widehat{V} = I^{-1} = - \left(\frac{\partial^2 \log L(\theta|Y, X)}{\partial\theta\partial\theta^T} \right)^{-1},$$

where L is the likelihood function and I is called the observed information matrix. The estimate of σ^2 is obtained using the residuals,

$$\widehat{\sigma}^2 = \left[\sum_i (y_i - \widehat{\mu}(x_i))^2 / V(\widehat{\mu}(x_i)) \right] / (n - p - 1).$$

When the sample size is large, $C(\theta - \widehat{\theta})$ is approximately multivariate normal with mean 0 and identity covariance matrix where $CC^T = \widehat{V}^{-1}$.

The likelihood ratio test is typically used to compare the effect of a block of covariates. As in the previous chapter, suppose that $X = (Z_1, Z_2)$, the full model has both Z_1 and Z_2 as predictors and the reduced model has only Z_2. Let L_F be the likelihood function from the full model evaluated at estimate $\widehat{\theta}_F$. Similarly, let L_R be the likelihood function from the reduced model evaluated at the estimate $\widehat{\theta}_R$. Note that parameters corresponding to Z_1 are set to zero in the reduced model. It can be shown that

$$LR = -2(\log L_R - \log L_F)$$

has a chi-square distribution with p_1 degrees of freedom where p_1 is the number of regressors in Z_1. For a quasi-likelihood model, there is no likelihood function. The generalization of the above procedure is performed using the deviance function (which reduces to a likelihood ratio test when appropriate).

4.2   Multiple Imputation Analysis

This section illustrates the many aspects of regression analysis using generalized linear models. Three models (logistic, Poisson and multinomial logistic) are considered here but other models can be fit using a similar approach. As in the case of linear models, the Jackknife Repeated Replication method can be used to estimate the completed data covariance matrix by treating the variance specification as the "working" variance function and not unduly affect the inferences.

4.2.1 Logistic Model

For the logistic regression example, we include a step-by-step presentation and discussion of the syntax used plus interpretation of results. For the remaining examples though, we show all syntax first and then turn to results and interpretation.

The logistic regression example uses the Primary Cardiac Arrest data set (stored in a working data set called "test") to predict a binary outcome of primary cardiac arrest (coded 1=Yes, 2=No). SAS with *IVEware* is used in this example. This data set includes study controls and cases, therefore imputation is done within each group separately with the IMPUTE command. Each imputation produces 5 complete data sets which are, in turn, combined into one stacked data set for subsequent analysis. The REGRESS command with a LINK LOGISTIC statement requests logistic regression. The logistic model predicts the probability of having cardiac arrest by age, gender and red-blood cell EPA/DHA values. Predicted estimates based on model predictions for specific types of individuals are also demonstrated. And, a joint test of age and gender is performed using PROC SURVEYLOGISTIC and PROC MIANALYZE. The test process used is similar to that presented previously for linear regression.

4.2.1.1 Imputation

```
<sas name="Logistic Regression Using Primary Cardiac
 Arrest Data">
/* Set libname */
libname pca 'P:\Ive_and_MI_Applications_Book\DataSets\PCA
and Omega 3 Fatty Acids Data';

data pca ;
set pca.test ;
* create variables for use in logistic regression ;
cardiac_arrest=2 ;
if casecnt=1 then cardiac_arrest=1 ; * omit no PCA in model
(highest category) ;
run ;

data case ;
 set pca ;
 if cardiac_arrest=1 ;
run ;

data control ;
 set pca;
 if cardiac_arrest=2 ;
run ;
```

```
/* Impute Missing Data For Cases */
<impute name="imputecase" >
title "Impute Missing Data" ;
datain   case ;
dataout imputecase all;
default continuous ; *age numcig yrssmoke fatindex
 dha_epa
redtot wgtkg totlkcal hgtcm;
transfer studyid casecnt ;
categorical casecnt race3 hyper diab smoke
fammi edusubj3
 cholesth gender;
mixed cafftot alcohol3 ;
restrict numcig(smoke=2,3) yrssmoke(smoke=2,3) ;
bounds numcig(>0) yrssmoke(>0,<=age-12)
 dha_epa(>0) redtot(>0)
 cafftot(>0) wgtkg(>0) totlkcal(>0) alcohol3(>0);
minrsqd .01;
iterations 5 ;
multiples 5;
seed 666 ;
run ;
</impute>

/* Impute Missing Data For Controls */
<impute name="imputecontrol" >
title "Impute Missing Data" ;
datain   control;
dataout imputecontrol all;
default continuous ; *  age numcig yrssmoke fatindex
 dha_epa redtot wgtkg totlkcal hgtcm ;
transfer studyid casecnt ;
categorical casecnt race3 hyper diab smoke fammi
edusubj3 cholesth gender;
mixed cafftot alcohol3 ;
restrict numcig(smoke=2,3) yrssmoke(smoke=2,3) ;
bounds numcig(>0) yrssmoke(>0,<=age-12) dha_epa(>0)
 redtot(>0)
 cafftot(>0) wgtkg(>0) totlkcal(>0) alcohol3(>0);
minrsqd .01;
iterations 5 ;
multiples 5;
seed 666 ;
run ;
```

```
</impute>

/* Combine Case and Control Data Sets*/
data all_imputations ;
 set imputecase imputecontrol ;
proc sort ;
by _mult_ ;
run;
proc freq ;
tables _mult_ ;
run ;
```

The preceding commands use the *IVEware* IMPUTE command to impute missing data separately within the case and control data sets. After imputation, the data sets are combined and sorted by the internal multiple imputation indicator variable called " _MULT_". These steps are needed to produce a working data set for regression analysis.

4.2.1.2   Parameter Estimates

```
/* Logistic Regression Using Imputed Data Sets */
<regress name="Logistic Regression PCA Regressed on
 Age Gender and RedBlood Cell EPA and DHA" >
title  "Logistic Regression Cardiac Arrest is Outcome" ;
datain all_imputations ;
link logistic ;
categorical gender ;
dependent cardiac_arrest ;
predictor age gender redtot ;
estimates Age35_fem_redtot7: intercpt(0), age(35),
gender(1), redtot(7.0) /
age35_male_redtot7: intercpt(0), age(35), gender(0),
redtot(7.0) ;
run ;
</regress>
```

Next, the above commands use the REGRESS command with a LINK LOGISTIC option to perform logistic regression to predict having a cardiac arrest event by age, gender, and EPA/DHA red blood cell counts, using the imputed data sets from the imputation process.

Table 4.1 below includes regression estimates in the form of Odds Ratios and 95% Confidence Limits.

Based on the logistic regression results, all else being equal, males are significantly more likely than women to have primary cardiac arrest, and those with higher levels of EPA/DHA have significantly lower odds of experiencing a PCA event.

Table 4.1: Results from Logistic Regression of Primary Cardiac Arrest

Variable	Odds Ratio	Confidence Interval
Age	1.017	(1.011, 1.023)
Gender	1.190	(1.024, 1.384)
Red Blood Cell EPA/DHA Counts	0.730	(0.691, 0.772)

4.2.1.3   Testing for Block of Covariates

```
/* data management and multivariate test using SAS */
data all_imp_sas ;
 set all_imputations ;
_imputation_=_mult_ ;
if gender=1 then female=1 ; else female=0;
proc sort ;
by _imputation_ ;
run ;
proc logistic data=all_imp_sas ;
by _imputation_ ;
model cardiac_arrest (event='1')=age female redtot / covb ;
ods output parameterestimates=outest covb=outcovb ;
run ;

proc mianalyze parms=outest covb=outcovb ;
modeleffects intercept age female redtot ;
testage_female: test age,female / mult ;
run ;
```

`</sas>`

Testing of a block or set of covariates is demonstrated with SAS PROC LOGISTIC AND PROC MIANALYZE. The above commands first perform needed data management to create a SAS data set suitable for use with PROC LOGISTIC and PROC MIANALYZE. Then, PROC MIANALYZE combines results from the logistic regressions and uses multivariate testing capabilities via the TEST statement with a /MULT option.

The output from PROC MIANALYZE in Table 4.2 includes a multivariate test of zero contribution to the logistic model for 2 variables, age and female. In this case, these variables have a significant contribution to the overall model with $F=3.18$, 2 df, and a p value of 0.042.

4.2.1.4   Estimate command

The ESTIMATE command syntax presented in Section 4.2.1.2 requests predicted estimates for men and women age 35 with red blood cell EPA/DHA

Table 4.2: Results from Multivariate Test of Age and Gender

Average Relative Increase in Variance	Num DF	Den DF	F	Pr>F
0.008	2	38745	3.18	0.042

count equal to 7.0. The results are presented in Table 4.3 and reveal small differences in estimates between men and women age 35 with the same red blood cell count of 7.0.

Table 4.3: Estimates for Age 35 Adults by Gender and Red Blood Cell Counts

Name	Estimate	Standard Error
Age 35 Females with Red Blood Cell Count=7	-1.440	0.230
Age 35 Males with Red Blood Cell Count=7	-1.615	0.213

4.2.2 Poisson Model

The Poisson model example is demonstrated with *IVEware* and R, using Health and Retirement Survey (HRS) 2012 data. The analytic goal is to predict the number of falls during the past two years, among those age 65+, with common health conditions such as back pain, depression, dementia, heart problems, cancer, and gender as covariates.

4.2.2.1 Full Code

```
<R name="Chapter_6_Poisson_Regression">
# note: settings file uses R.exe command to invoke R
# set working directory to correct path
setwd("P:/ive_and_MI_Applications_Book/Chapter6GenRegression
/Poisson Regression")
(WD <- getwd())
# import the input dataset
hrs <- read.table("poisson_hrs_small_25may2016.txt",
sep="\t", header=TRUE)
save(hrs, file="hrs.rda")
summary(hrs)
# run iveware
# multiple imputation
# Impute Missing Data
<impute name="impute_mult1">
title Poisson_Regression_Example_Chapter6_HRS_2012;
datain  hrs ;
dataout impute_mult1 ;
```

```
default categorical ;
continuous nage nwgtr ;
count numfalls24 ;
transfer hhid pn ;
multiples 5 ;
iterations 10 ;
seed 425;
run ;
</impute>

# extract the remaining four multiply imputed datasets
<putdata name="impute_mult1" mult="2" dataout="impute_mult2" />
<putdata name="impute_mult1" mult="3" dataout="impute_mult3" />
<putdata name="impute_mult1" mult="4" dataout="impute_mult4" />
<putdata name="impute_mult1" mult="5" dataout="impute_mult5" />

# run Poisson Regression
<regress name="C6_Poisson_Regression">
title  Chapter 6 Poisson Regression ;
datain impute_mult1 impute_mult2 impute_mult3 impute_mult4
impute_mult5;
stratum stratum ;
cluster secu ;
weight  nwgtr ;
categorical gender backpain depress dementia heartcondition
cancer ;
dependent numfalls24 ;
predictor gender backpain depress dementia heartcondition
cancer ;
link log ;
run ;
</regress>
</R>
```

The code first imports a subset of 2012 HRS data into R (with the read.table command) and saves an R format data object called "hrs". IMPUTE is used to impute missing data and output 5 imputed data sets with the PUTDATA command. Next, REGRESS with a LINK LOG statement is used to request Poisson regression. Note that the complex sample design features and weights are used to estimate "design-based" variances but an alternative method is the Bayesian Bootstrap implemented in the BBDESIGN command of *IVEware*. The remaining code statements are similar to those used in previous linear and logistic regression examples.

Based on the Poisson regression results in Table 4.4, as compared to women, males have significantly more estimated falls during the 2 year exposure period while each of the physical and mental health conditions have

Table 4.4: Results from Poisson Regression of Number of Falls During Past 2 Years

Variable	Risk Ratio	Confidence Interval
Gender	1.257	(1.081,1.462)
Back Pain	1.707	(1.512, 1.929)
Depression	1.728	(1.517, 1.969)
Dementia	2.668	(2.102, 3.387)
Heart Condition	1.447	(1.292, 1.621)
Cancer	1.137	(0.926, 1.396)

significantly higher estimated rates of falls, as compared to those without physical or mental health conditions, holding all else equal.

4.2.3 Multinomial Logit Model

The third example demonstrates multinomial logistic regression using NCS-R data. This example presents use of Stata with *IVEware*.

Variables from the Personality Section from Part 2 of the NCS-R survey were collapsed into variables representing personality traits including "loner", "inflexible", "socially awkward", and "suspicious". These traits plus gender are used to predict a nominal outcome variable representing employment status (coded 1=Employed, 2=Previously Employed, 3=Out of the Labor Force).

4.2.3.1 Full Code

```
<stata name="Multinomial_Logistic_Regression_Using_NCSR_Data">
cd "P:\ ive_and_MI_Applications_Book\Chapter6GenRegression
\Multinomial Regression"
use ncsr_pea_sub_22apr2016

/* Impute Missing Data For PEA Categorical Variables*/
<impute name="impute" >
title Impute Missing Data;
datain  ncsr_pea_sub_22apr2016;
dataout impute all;
default categorical;
transfer SampleID ;
continuous finalp2wt Age bmi ;
multiples 5;
iterations 10 ;
seed 222 ;
diagnose pea76 ;
run ;
</impute>
```

```
/* generate dummy variables from imputed PEA variables*/
gen loner=2
replace loner=1 if PEA76==1 | PEA81==1
gen inflexible=2
replace inflexible=1 if PEA77==1 | PEA78==1
gen awkward=2
replace awkward=1 if PEA79==1 | PEA80==1
gen suspicious=2
replace suspicious=1 if PEA82==1 | PEA83==1
gen female=2
replace female=1 if sexf==1

* reverse coding of employment category for ease of
interpretation
gen rev_empcat=.
replace rev_empcat=3 if empcat==1
replace rev_empcat=2 if empcat==2
replace rev_empcat=1 if empcat==3
save impute3, replace

<regress name="Multinomial_Logistic">
title  Multinomial Logistic Regression ;
datain impute3 ;
link logistic ;
categorical rev_empcat loner inflexible awkward
suspicious female ;
dependent rev_empcat ;
predictor female loner inflexible awkward suspicious;
weight finalp2wt;
stratum str ;
cluster secu ;
run ;
</regress>
</stata>
```

In this example, the imputation of missing data is performed with the IM-PUTE command before final variable construction and subsequent analysis of completed data sets. This allows variable recoding and construction to be done with imputed raw variables. Next, five completed data sets are output as one stacked file by the ALL option on the DATAOUT statement. This option creates a data set with the 5 multiples vertically stacked and identified by the _MULT_ variable. After imputation, a series of dummy variables are created with reverse coding, as needed, to assist in interpretation for selected variables. Note that all data management steps are done in Stata.

Finally, the REGRESS command is used for polytomous or multinomial logistic regression. The LINK LOGISTIC statement requests logistic regression and the CATEGORICAL statement declares the dependent variable and independent variables to be treated as categorical. The dependent variable in this example has 3 discrete values therefore multinomial regression is appropriate. The additional REGRESS statements are similar to previous examples.

Table 4.5: Results from Multinomial Logistic Regression of Employment Status

Variable	Odds Ratio	Confidence Interval
Outcome=Employed		
Female	1.438	(1.086, 1.903)
Loner	1.491	(1.166, 1.906)
Inflexible	0.689	(0.567, 0.837)
Awkward	0.723	(0.583, 0.898)
Suspicious	1.08	(0.847, 1.390)
Outcome=Previously Employed		
Female	2.389	(1.865, 3.060)
Loner	1.356	(1.103, 1.666)
Inflexible	0.912	(0.759, 1.096)
Awkward	1.446	(1.140, 1.835)
Suspicious	1.114	(0.907, 1.367)

Based on Table 4.5, in the first model (Outcome=Employed), women and loners are significantly more likely than men or non-loners to be employed versus out of the labor force (OOLF) while those that are inflexible, awkward are significantly less likely than those not in these groups to be employed v. OOLF. And, suspicious respondents are non-significantly more likely to be employed v. OOLF, as compared to those not considered suspicious.

Based on results from the second model (Outcome=Previously Employed), compared to men or those not loners or awkward, women, loners, and the awkward are significantly more likely to be previously employed rather than out of the labor force. Compared to their reference groups of not in the group of interest, those inflexible are less likely (non-significant) to be previously employed v. OOLF and those considered suspicious are slightly more likely (non-significant) to be previously employed v. OOLF.

4.3 Additional Reading

In addition to the regression references provided in Chapter 5, a comprehensive review of regression analysis with missing predictors is given in Little

(1992). For logistic regression and extensions, see Agresti (2012) and for Poisson regression and count models see Long and Freese (2006) and Long (1997). See also O'Neill and Temple (2012).

4.4   Exercises

1. Download the St. Louis Risk Research Project data from the book web site and examine the contents of the data set. The analysis goal is to perform logistic regression using the binary variable representing high adverse symptoms (average per family, coded yes or no) predicted by moderate and high parental risk scores. However, missing data is an issue in this data set and requires imputation.

   (a) Use your software of choice to evaluate the missing data problem. How much missing data exists on each variable? What variable(s) does not have any missing data?

   (b) Use IMPUTE to impute missing data and create 25 (with 5 iterations) data sets. Make sure to declare categorical variables correctly in the imputation, use the ALL statement to create a "long" output data set, and use a seed value of 2016. What assumption are you making if you impute all groups of parental risk together? What is the rationale for 25 multiples?

   (c) Check the quality of the imputations informally by studying the IMPUTE output and also by performing a means analysis by multiple for each imputed variable (using your chosen software). What are the key parts of the output to check? Do you see any issues that might require a "fix"? For a more formal approach for imputation diagnostics, use DIAGNOSE with the /ALL option for one imputed variable of your choice. Does this reveal any problems or are the results similar to the informal approach?

   (d) Using the long file containing the imputed data sets, create an indicator of high number of child symptoms per family called HIGH_SYMPTOM (coded as 1 if imputed s1=2 and s2=2, else coded 0). Also, create 2 binary indicator variables using the G variable where MODRISK=1 if G=2, else MODRISK=0 and HIGHRISK=1 if G=3, else HIGHRISK=0, (omit G=1, Low Risk). Next, use the REGRESS command to perform logistic regression with the outcome HIGH_SYMPTOM predicted by MODRISK and HIGHRISK. How does the REGRESS command incorporate the multiple imputation variability in the analysis?

(e) Provide a brief paragraph interpreting the logistic regression results. Does parental risk group have a significant impact on a high number of adverse symptoms among children?

2. Download the data set called **EX_POISSON_EXERCISE_HRS** (a subset of 2012 HRS data) from the book web site. This exercise asks you to impute missing data on key variables and use Poisson regression to predict number of days respondents drank in an average week by common demographic and health conditions.

   (a) Prepare for imputation through examination of the variable contents, type, and amount of missing data in the data set. How much missing data exists and how would you describe the missing data pattern? Make sure to set any variables where values of 8 (DK) or 9 (Refused) remain as "valid values" to missing prior to the imputation. Based on this work, how many multiples do you recommend and how do you justify this decision?

   (b) Impute missing data with IMPUTE. Make sure to correctly declare each variable as CATEGORICAL, CONTINUOUS, TRANSFER, MIXED, or COUNT. (Hint: set HHID and PN to TRANSFER and set the count of days drink per week to type COUNT). Include bounds for any imputed continuous variables and a restriction so that NC129 is imputed only for those that drink (NC128=1). Use a SEED value of 2016. Why are bounds and restrictions required in this imputation?

   (c) Examine the output from the IMPUTE process and check for any imputed values outside of the observed bounds, imputed v. observed distributions that are very different, and any double-counted imputed variables. Do you see any problems with the imputations? If so, what would you do to correct the issues?

   (d) Run a Poisson regression using the imputed data sets with the REGRESS command with a LINK LOG option and STRATUM, SECU, and WEIGHT statements to obtain a design-based and MI Poisson regression. The outcome variable is imputed number of days drink alcohol in an average week (NC129, range from 0 to 7) predicted by the variables GENDER, MARCAT, DIABETES, and ARTHRITIS. Make sure to treat GENDER and MARCAT as categorical and DIABETES and ARTHRITIS as binary indicators in the model. As a reminder, the highest category of categorical variables is omitted by default while indicators coded as 1/0 will be handled as having the condition of interest, say arthritis, versus not having the condition of interest.

   (e) Write a short summary of the results and explain how the complex sample design features and imputation variability are incorporated in this analysis. What covariates have a significant impact on the number of days drink alcohol in an average week?

3. Download the **EX_MTF_2014** data set from the book web site and translate to a format useable for your software of choice. The goal of this exercise is use of multinomial logistic regression to predict type of high school attended by gender, race, and if ever smoked Marijuana. The data set is a subset of 2014 Monitoring the Future data for high school seniors (Monitoring the Future: A Continuing Study of American Youth (12th-Grade Survey), 2014 (ICPSR 36263).

(a) Prepare for imputation through data exploration and determine variable type, extent of missing data, and meaning of variable values. What variables have missing data or have values that may require re-coding prior to analysis? Which have the highest rates of missing data? What is the distribution of the dependent variable, v2172, and what does each value mean? Create a new version of the dependent variable called HSTYPE where .=. (Missing), 1=1 (College Prep), 2=2 (General HS), and 3 and 4 are collapsed into 3=Voc/Tech/Other. Also, create a series of indicator variables named FEMALE, BLACK, WHITE, HISPANIC, and ASIAN/OTHER by coding each =1 if in the specified group and 0 otherwise.

(b) Impute missing data with the IMPUTE command and create M=15 multiples. Treat CASEID as a TRANSFER variable, ARCHIVE_WT as a CONTINUOUS variable, set the default to CATEGORICAL, and SEED equal to 2014. Also, use the DIAGNOSE option with both HSTYPE and v2150.

(c) Examine the imputation output and plots from DIAGNOSE to identify possible issues with the imputation. Do you see any issues to investigate? If so, describe the process you might follow to resolve the apparent problems.

(d) Run an un-weighted complete case multinomial logistic regression using the original data as follows: HSTYPE regressed on FEMALE, WHITE, HISPANIC, ASIAN/OTHER and V2115D.

(e) Next, repeat the analysis in part d. but use the 15 multiply imputed data sets. Compare the results and describe how multiple imputation of missing data improves the analysis. How many observations are used in the complete case analysis versus those used in the MI analysis?

(f) Repeat parts d. and e. but use the complex sample design "pseudo" variables (created as below) and weight variable in the analysis. To create the pseudo design variables, first create a new variable called STRATUM which is set equal to 1 for each record in the data set and a second variable called CLUSTER set equal to the CASEID variable. Use the STRATUM, CLUSTER, and WEIGHT statements in the complete case and MI analyses. Does weighting and incorporating the pseudo complex design variables make a difference in your overall conclusions? (Note that processing time will be much longer for these models due to use of the Jackknife Repeated Replication method for variance estimation).

5

Categorical Data Analysis

5.1 Contingency Table Analysis

Regression models with a categorical outcome variable were discussed in Chapter 4. Many categorical data analysis problems, however, involve assessing association between several categorical variables without specifying one as an outcome variable and others as predictors or independent variables. The simplest analysis is assessing association between two categorical variables arranged as a two-way contingency table. Suppose Y_1 and Y_2 are two categorical variables with Y_1 taking R possible values labeled $r = 1, 2, \ldots, R$ and Y_2 taking C possible values labeled $c = 1, 2, \ldots, C$. Let $\pi_{rc} = Pr(Y_1 = r, Y_2 = c)$ define the joint distribution of (Y_1, Y_2). Let $\pi_{r+} = Pr(Y_1 = r)$ and $\pi_{+c} = Pr(Y_2 = c)$ be the marginal probabilities.

Based on the sample of size n, let n_{rc} be the number of subjects in cell (r, c) in a table with R rows and C columns. Let $n_{r+} = \sum_c n_{rc}$ and $n_{+c} = \sum_r n_{rc}$ be the marginal frequencies leading to $\hat{\pi}_{rc} = n_{rc}/n$, $\hat{\pi}_{r+} = n_{r+}/n$ and $\hat{\pi}_{+c} = n_{+c}/n$. These are the maximum likelihood estimates assuming that n_{rc} follows a multinomial distribution with cell probabilities π_{rc} where $\sum_{rc} \pi_{rc} = 1$. Additional constraints needed are $\sum_r \pi_{r+} = 1$ as well as $\sum_c \pi_{+c} = 1$. Thus the cell probabilities need to satisfy $(R - 1) + (C - 1) + 1$ constraints leading $R \times C - \{(R - 1) + (C - 1) + 1\} = R \times C - R - C + 1 = (R - 1) \times (C - 1)$ free parameters. Though there are $R \times C$ cell frequencies, there is only $(R - 1) \times (C - 1)$ degrees of freedom because the estimates will also have to satisfy the constraints. The multinomial distribution arises from assuming a Poisson distribution for the individual cell frequencies and conditioning on the observed sample size and the row and column margins.

The goal of the analysis is to assess whether the two variables are independent. Under the independence assumption, $\pi_{rc} = \pi_{r+}\pi_{+c}$, and one would expect that relationship to hold in the estimates as well. Thus, the distance between n_{rc}/n and $n_{r+}/n \times n_{+c}/n$ can be used as a measure of lack of fit of the data to the hypothesis (of independence). A chi-square statistic measuring this distance is defined as

$$D = \sum_r \sum_c \frac{(n_{rc} - n_{r+}n_{+c}/n)^2}{n_{r+}n_{+c}/n}$$

$$= \sum_r \sum_c \frac{(n_{rc} - e_{rc})^2}{e_{rc}}$$

where $e_{rc} = n_{r+}n_{+c}/n$ is called the (estimated) expected frequency under the independence assumption. Under the independence assumption, the statistic D has a chi-square distribution with $(R-1) \times (C-1)$ degrees of freedom. Thus, if the observed value of D is in the upper tail of the chi-square distribution then the hypothesis is a suspect. (Interestingly, R. A. Fisher reanalyzed Gregor Mendel's data and concluded that the chi-square statistic was too small to be plausible and, hence, the data should be suspect. Thus, it may be important to look at the both tails based on the calculated value of the chi-square statistic.)

Missing values in either variable will result in a partially classified table. Subjects with the observed values of (Y_1, Y_2) yield a cross-classified table, subjects only with the observed values of Y_2 provide information about its marginal distribution and those with only Y_1 observed provide information about its marginal distribution. Suppose that the missing values have been multiply imputed resulting in M completed-data sets and $d_l, l = 1, 2, \ldots, M$ are the completed data chi-square statistics. The multiple imputation chi-square statistic is computed as follows. Let $\bar{d}_{MI} = \sum_l d_l/M$, $\bar{p}_{MI} = \sum_l \sqrt{d_l}/M$ and $v_{MI} = (1 + 1/M)\sum_l(\sqrt{d_l} - \bar{p}_{MI})^2/(M-1)$. Define,

$$\tilde{D}_{MI} = \frac{\bar{d}_{MI}/\{(R-1)(C-1)\} - (M+1)v_{MI}/(M-1)}{1 + v_{MI}}$$

and is referred to an F-distribution with $(R-1)(C-1)$ as the numerator degrees of freedom and $\nu_d = \{(R-1)(C-1)\}^{-3/M}(M-1)(1+v_{MI}^{-2})$ as the denominator degrees of freedom.

5.2   Log-linear Models

When more than two variables are involved, the association between them is expressed using a log-linear model for the expected cell counts (or, equivalently, the cell probabilities). For example, in the $R \times C$ contingency example discussed in the previous section, the expected cell count for the cell (r, c) is $\mu_{rc} = n \times \pi_{rc}$. A general log-linear model takes the form,

$$\log \mu_{rc} = \mu + \lambda_r^{(1)} + \lambda_c^{(2)} + \lambda_{rc}^{(12)}$$

and the constraints can be expressed by requiring $\sum_r \lambda_r^{(1)} = \sum_c \lambda_c^{(2)} = \sum_r \sum_c \lambda_{rc}^{(12)} = 0$. These constraints ensure that $\sum_r \sum_c \mu_{rc} = n$. The log-linear model is similar to the ANOVA type structure for the logarithm of the cell frequencies. The parameter μ is the overall average cell count, $\lambda^{(1)}$ is the main effect of Y_1, $\lambda^{(2)}$ is the main effect of Y_2 and $\lambda^{(12)}$ represents the interaction between the two variables. This is called the fully saturated model as no structure is placed on the cell probabilities (or frequencies). This model is a restructure of the RC cell probabilities representing the joint distribution of (Y_1, Y_2). The number of parameters used in this representation is μ, $R-1$ values of $\lambda^{(1)}$, $(C-1)$ values of $\lambda^{(2)}$ and $(R-1)(C-1)$ values of $\lambda^{(12)}$. That is, $\nu_S = 1 + (R-1) + (C-1) + (R-1)(C-1) = RC$ (with the constraint that $\sum_r \sum_c \mu_{rc} = n$). This is a perfect fit for the two-way table (as many parameters as the number of cells), hence, is called the saturated model.

Under this setup, independence between Y_1 and Y_2, is equivalent to setting the interaction terms $\lambda_{rc}^{(12)} = 0 \ \forall \ r,c$. Sometimes, a short hand notation, $[Y_1][Y_2]$, is used to represent this model (the saturated model is $[Y_1 Y_2]$). The reduced model is

$$\log \mu_{rc} = \mu + \lambda_r^{(1)} + \lambda_c^{(2)}.$$

The number of parameters in this model is $\nu_I = 1 + (R-1) + (C-1) = R+C-1$. The saturated and independence models are usually compared using the likelihood ratio test. Let l_S and l_I be the log-likelihoods from the saturated and independence model, respectively. The likelihood ratio test statistic is $G = 2(l_S - l_I)$ which has a chi-square distribution (under the independence model) with $\nu = \nu_S - \nu_I = (R-1)(C-1)$ degrees of freedom.

The attractive feature of the log-linear model is that many parameters are easily interpretable. Consider the case where $R = C = 2$. The following table represents the logarithm of the cell frequencies in terms of the parameters:

Cell	$C=1$	$C=2$	Total
$R=1$	$\mu + \lambda_1^{(1)} + \lambda_1^{(2)} + \lambda_{11}^{(12)}$	$\mu + \lambda_1^{(1)} + \lambda_2^{(2)} + \lambda_{12}^{(12)}$	$2\mu + 2\lambda_1^{(1)}$
$R=2$	$\mu + \lambda_2^{(1)} + \lambda_1^{(2)} + \lambda_{21}^{(12)}$	$\mu + \lambda_2^{(1)} + \lambda_2^{(2)} + \lambda_{22}^{(12)}$	$2\mu + 2\lambda_2^{(1)}$
Total	$2\mu + 2\lambda_1^{(2)}$	$2\mu + 2\lambda_2^{(2)}$	4μ

The log odds ratio is

$$\log \left(\frac{\mu_{11}\mu_{22}}{\mu_{12}\mu_{21}} \right) = \lambda_{11}^{12} + \lambda_{22}^{(12)} - \lambda_{12}^{(12)} - \lambda_{21}^{(12)}.$$

The marginal log odds ratio for Y_1 is $\log(\mu_{2+}/\mu_{1+}) = 2(\lambda_2^{(1)} - \lambda_1^{(1)})$ and the marginal log-odds ratio for Y_2 is $\log(\mu_{+2}/\mu_{+1}) = 2(\lambda_2^{(2)} - \lambda_1^{(2)})$.

Thus, the parameter estimates, and specifically the contrasts among the parameters provide useful quantitative measures of strengths of associations or the effects of variables. The contrast specifications are very similar to those used in the analysis of variance of a continuous variable with several categorical predictors. Here the cell counts play the role of a continuous variable with categorical variables as predictors. This similarity has been used (see Grizzle, Starmer and Koch (1969)) to develop a Weighted Least Squares (WLS) approach for the analysis of contingency table using a normal approximation for the Poisson distribution.

As before let, $\pi_{rc}, r = 1, 2, \ldots, R; c = 1, 2, \ldots, C$ denote the cell probabilities and p_{rc} be the corresponding estimates from the sample data. The goal is to fit a linear model,

$$F(\pi) = X\beta$$

where F is a transformation determined by substantive interest and X are covariates or the model matrix.

For large samples, using the Taylor series approximation, $F(p) \approx F(\pi) + G(p - \pi)$, we have $E(F(p)) \approx F(\pi)$ and $V_F = Var(F(p)) = GVG^T$ where V is the covariance matrix of the sample proportions p, and G is the first derivative of the function F evaluated at the sample proportions p.

Thus, the general linear model framework can be used to fit the model

$$F(p) = X\beta + \epsilon$$

where $\epsilon \sim N(0, V_F)$.

The weighted least squares estimate of β is $\widehat{\beta} = (X^T V_F^{-1} X)^{-1} X^T V_F^{-1} F(p)$ and its covariance matrix is $(X^T V_F^{-1} X)^{-1}$. Defining $F(\pi) = \log \pi$ provides the WLS analog of the log-linear model. The ANOVA table, break down of various sums of squares using contrasts and partial F-tests for comparing models, and other tools used in the regression analysis can be used to infer about the main and interaction effects. All the approaches discussed previously in Chapter 3 can be used in the missing data context as well.

5.3   Three-way Contingency Table

Consider a third variable Y_3 taking the values $h = 1, 2, \ldots, H$. The saturated log-linear model can be expressed as

$$\log \mu_{rch} = \mu + \lambda_r^{(1)} + \lambda_c^{(2)} + \lambda_h^{(3)} +$$

$$\lambda_{rc}^{(12)} + \lambda_{rh}^{(13)} + \lambda_{ch}^{(23)} +$$

$$\lambda_{rch}^{(123)}$$

For a three-variable example, (Y_1, Y_2, Y_3), the notation $[Y_1][Y_2][Y_3]$ denotes the model where all of the three variables are mutually independent, and, thus, the log-linear model contains only the main effects (two-way and three-way interaction terms, $\lambda^{(12)}, \lambda^{(13)}, \lambda^{(23)}$ and $\lambda^{(123)}$ are set to 0).

The notation $[Y_1 Y_2][Y_3]$ implies that Y_1 and Y_2 are associated but this association is the same across the levels of Y_3 and the population distribution across the categories of Y_3 are not equal. Here the model contains all three main effects and the interaction between Y_1 and Y_2. Similarly,

$[Y_1 Y_2], [Y_1 Y_3], [Y_2 Y_3]$ denotes the model with all main effects, two way interaction effects but no three-way interaction terms. This implies that the association between Y_1 and Y_2 is the same across the levels of Y_3 (and the same is true for (Y_1, Y_3) across the levels of Y_2; and for (Y_2, Y_3) across the levels of Y_1).

5.4 Multiple Imputation

Consider a three-way contingency table with an arbitrary pattern of missing data. The completely observed units lead to an $R \times C \times H$ contingency table and subjects with missing values may provide information about the two-way and one-way marginal tables. Maximum Likelihood Estimates (MLE) can be obtained using iterative algorithms such as the EM-algorithm and specialized codes will have to be developed. The multiple imputation analysis might be more useful because one can incorporate auxiliary variables in the imputation process and the complete data software can be used to obtain estimates and then combine them to form a single inference.

The imputation of Y_1, Y_2 and Y_3 must ensure that the interaction effects are preserved, however small. Otherwise, the completed data analysis will tilt the estimates of the ignored interaction effects towards zero. Suppose that X denotes auxiliary variables that are correlated with Y_1, Y_2 and Y_3. The suggested imputation model for Y_i is a multinomial logit model with predictors $(Y_j, Y_k, Y_j \times Y_k, X, \ldots)$ where (i, j, k) is a permutation integers $(1, 2, 3)$ and $\ldots$ may include some interaction terms with X and Y. Similarly, for a p-way contingency table, include all interactions up to $(p - 1)$ in the imputation model. This may not be the most efficient approach because it errs on the side of the imputation model being as close to the saturated model as possible. An alternative is to trim the model by dropping interactions that may not be of interest (such as 4-way or higher order interactions terms as these are hard to interpret). Sensitivity of inferences can be explored by imputing the missing values under the general and reduced models.

5.5 Two-way Contingency Table

This example uses data from Table 13.1 of Little and Rubin (2002). The data was converted from tabular format with cell counts to a SAS data set using the data step and the macro language. Two 2 categorical variables are used in this example; Y1 coded as 1 or 2 and Y2 coded as 1,2, or 3. Due to missing data on Y2 (190 cases are missing on Y2 but observed on Y1), multiple imputation was performed.

SAS with *IVEware* is used in the demonstration. Imputation is performed with the IMPUTE command followed by a table analysis of Y1 and Y2 with the DESCRIBE command, using the concatenated imputed data sets. To generate a combined Chi-Square statistic, use of the SAS macro language and PROC IML is demonstrated (modified from a conference paper by Ratitch, Lipkovich, O'Kelly, 2013). And, for log-linear analysis, PROC CATMOD is executed for each imputation multiple with a PRED=PROB option used in the model statement to produce cell probabilities for each of the 6 levels of $Y1 \times Y2$. These results are combined using PROC MIANALYZE and represent MI adjusted cell probabilities for the cells from the two way contingency table.

```
/* Data Set Up and Imputation Code */

<sas name="Two Way Contingency Table">
/*macro to create output raw data for example*/
%macro it (y1, y2, c) ;

data cat ;
do i=1 to &c   ;
 y1=&y1 ;   y2=&y2   ;   count=&c ;
 output ;
end ;
drop i count ;
run ;

proc append base=catall data=cat ;
run ;

%mend ;
%it(1,1,20) ;
%it(1,2,30) ;
%it(1,3,40) ;
%it(2,1,50) ;
%it(2,2,60) ;
%it(2,3,20) ;
%it(1,. , 100) ;
%it(2,. , 90) ;

proc freq data=catall ;
 tables y1 y2 y1*y2 / missing ;
run ;

proc mi nimpute=0 data=catall;
```

```
  var y1 y2 ;
run ;

/* impute missing data */
<impute name="c5_ex1_impute">
  title "Categorical Example 1 Imputation for Two Way Table" ;
  datain catall ;
  dataout cat1imp all;
  default categorical;
  iterations 5;
  multiples 15;
  seed 67566;
run;
</impute>

/* table analysis with imputed data */
<describe name="c5_ex1_table">
 title "Categorical Example 1 Table Analysis of Imputed Data " ;
 datain cat1imp ;
 categorical y1 y2 ;
 table y1*y2 ;
 run ;
</describe>
```

The previous code block first creates a working data set using a user-defined macro and next, imputes missing data with IMPUTE (M=15) using typical code (similar to previous examples). This step is followed by a table analysis of Y1 and Y2 with the DESCRIBE command, using the concatenated imputed data sets.

Table 5.1: Cell Proportions from Two Way Contingency Table Analysis

Y1		Y2	
	1	2	3
1	0.10000	0.15789	0.20553
2	0.21154	0.24488	0.08016

Table 5.1 presents cell proportions for the two way contingency table produced by the DESCRIBE command.

5.5.1 Chi-square Analysis

```
/* ChiSq Pooling done in SAS macro, df=(r-1)(c-1) or
2 for 2 by 3 table */
```

```
* prepare output data sets from PROC FREQ ;
proc sort data=cat1imp ; by _mult_ ; run ;
proc freq data=cat1imp;
tables y1*y2 / chisq  ;
by _mult_ ;
ods output chisq=outchisq (where=(statistic eq 'Chi-Square'));
run ;
proc print data=outchisq ;
run ;


%macro computePooledCh(datain,dataout,df=1);
Proc iml;
 USE &datain ;
 READ ALL VAR {value} INTO chval;    * read values of chisq
 across the 15 imputations ;
 df=&df;
 m=NROW(chval);
 cvalroot_m = sum(chval##0.5)/m;
 cval_m = SUM(chval)/m;
 a=(chval##0.5-j(m,1,1)*cvalroot_m)##2;
 rx = sum(a)*(1+1/m)/(m-1);
 Dx=(cval_m/df - (m+1)/(m-1)*rx)/(1+rx);
 df_den=(df**(-3/m))*(m-1)*(1+1/rx)**2;
 Pval=1-CDF("F",Dx,df,df_den);

 create F from dx[colname={"DX"}] ;
 append from dx ;
 create df_den from df_den[colname={"DF_den"}] ;
 append from df_den ;
 create rx from rx[colname={"rx"}] ;
 append from rx ;

 create imputations from m[colname={"M"}] ;
 append from m ;
 CREATE mean_chisq FROM cval_m[colname={"MeanCHISQ"}] ;
 append from cval_m ;
 create df from df [colname={"DF"}] ;
 append from df ;
 create &dataout from Pval[colname={"PvalPooledCh"}] ;
 append from Pval ;

run ; quit ;
%mend;


/* call macro with 2 df for 2 by 3 table and create output
```

```
data set called pooledchisq */
options symbolgen mprint ;
%computepooledch(outchisq, pooledchisq,df=2) ;


/* data set from ChiSq macro output */
data chisq_pval ;
 merge imputations f rx df_den mean_chisq df pooledchisq ;
proc print ;
run ;
```

The above code demonstrates preparation of a combined chi-square statistic for the two way contingency table generated from an output data set from PROC FREQ with a CHISQ option on the tables statement. The combined chi-square is ultimately obtained from PROC IML code embedded in the SAS user-defined macro called computepooledch. The results are presented in Table 5.2 and suggest a significant association between Y1 and Y2 (F=12.220, numerator df=2, denonminator df=61.856, and p=0.000).

Table 5.2: Results from Chi-Square Analysis using PROC IML

F-statistic	Numerator DF	Denominator DF	p
12.22	2	61.856	0.000

5.5.2 Log-linear Model Analysis

```
/* Log Linear Model for Two Way Table with r=2 and c=3 Table */
/* PROC CATMOD with PROC MIANALYZE*/
data cat1imp1 ;
 set cat1imp ;
 _imputation_=_mult_ ;
run ;
proc sort ;
 by _imputation_ ;
run ;

proc catmod data=cat1imp1 ;
 by _imputation_ ;
 model y1*y2 =_response_  / noparm pred=prob ;
 loglin y1 y2 y1*y2 ;
 ods output predictedprobs=outprobs ;
run;

/*print out outprobs data set*/
proc print data=outprobs ;
```

```
run ;
/*PROC MIANALYZE to combine predicted prob and SE */
proc sort ; by functionnum _imputation_ ; run ;

proc mianalyze data=outprobs ;
 by functionnum ;
 modeleffects predfunction ;
 stderr predstderr ;
run ;
</sas>
```

The above command syntax employs PROC CATMOD and PROC MI-ANALYZE to perform a log-linear analysis of the 15 completed data sets. The PROC CATMOD setup requests response probabilities for Y1, Y2, and Y1*Y2, that is, specifies a log-linear model of main effects plus the interaction of Y1*Y2. It also requests cell probabilities (PRED=PROB) saved in an output data set appropriate for PROC MIANALYZE. The MIANALYZE procedure is used to combine predicted cell probabilities from the log-linear model. Because this is a saturated model, the cell proportions match those of the table analysis results presented in Table 5.2.

Table 5.3: Cell Proportions from Log-linear Analysis

Y1		Y2	
	1	2	3
1	0.10000	0.15789	0.20553
2	0.21154	0.24488	0.08016

5.6 Three-way Contingency Table

5.6.1 Log-linear Model

The next example uses data from Table 13.8 from Little and Rubin (2002), originally analyzed by Bishop, Fienberg and Holland (1975, Table 1.4-2). There are 3 variables of interest: Clinic, coded as 1=A or 2=B, Prenatal Care, coded 1=Less and 2=More, and Survival, coded 1=Died and 2=Survived. The main analytic goal is to use a Log-linear model to analyze the relationships between the 3 categorical variables. However, 255 cases are missing data on type of Clinic while n=715 cases are fully observed on each variable. Prior to analysis, imputation of missing data is addressed by use of IMPUTE. This is followed by analysis of completed data sets using PROC CATMOD and PROC MIANALYZE using SAS.

The *IVEware*/SAS code is presented in full before results in this example. The first section of code prepares a SAS data set from tabular information with frequency counts, imputes missing data on the clinic variable with interaction terms in the imputation models, and produces 15 imputation multiples. Next, PROC CATMOD is used to execute 3 log-linear models using the 15 completed data sets: one model is fully saturated and 2 additional models with a set of 2 way interactions are considered. Finally, use of PROC MIANALYZE to combine predicted cell proportions from CATMOD step is demonstrated.

```
<sas name="Three Way Contingency Table">
/*Data creation and imputation*/

%macro it1 (c, p, s, count) ;
data cat2 ;
do i=1 to &count ;
c=&c ;  p=&p; s=&s ; count=&count ;
output ;
end ;
drop i count ;
run ;
proc append base=catall2 data=cat2 ;
proc print ;
run ;

%mend ;
%it1(1,1,1, 3) ;
%it1(1,1,2,176) ;
%it1(1,2,1,4) ;
%it1(1,2,2,293) ;

%it1(2,1,1, 17) ;
%it1(2,1,2,197) ;
%it1(2,2,1, 2) ;
%it1(2,2,2,23) ;

%it1(.,1, 1, 10) ;
%it1(.,1, 2, 150) ;
%it1(.,2,1,5) ;
%it1(.,2,2,90) ;

proc print data=catall2 ;
run ;
proc format ; value cf 1='A' 2='B' ; value pf
1='Less' 2='More' ; value sf 1='Died' 2='Survived' ;

proc freq data=catall2 ;
```

```
  tables c*p*s / missing list   ;
  format c cf. p pf. s sf. ;
run ;

proc mi nimpute=0 data=catall2 ;
  var c p s   ;
run ;

/* impute missing data */

<impute name="Ex2_impute" >
title "Categorical Example 2 Imputation for Three Way Table" ;
  datain catall2 ;
  dataout cat2imp all;
  default categorical;
  interact c*p c*s p*s ;
  iterations 5;
  multiples 15;
  seed 67566;
run;
</impute>

<describe name="Ex2_Table_Analysis">
title "Categorical Example 2 Table Analysis of Imputed Data " ;
 datain cat2imp ;
 categorical c p s   ;
 table c*p*s ;
 run ;
  </describe>

/* Run log-linear models*/

/*First model is fully saturated*/
data cat2imp1 ;
 set cat2imp ;
 _imputation_=_mult_ ;
run ;

proc sort ; by _imputation_ ; run ;

proc catmod data=cat2imp1 ;
 by _imputation_ ;
 model c*p*s =_response_  / noparm pred=prob ;
 loglin c|p|s ;
 ods output predictedprobs=outprobs_sat ;
```

```
run;

/*print out outprobs data set*/
proc print data=outprobs_sat ;
run ;

/*PROC MIANALYZE to combine predicted prob and SE */
proc sort ; by functionnum _imputation_ ; run ;

proc mianalyze data=outprobs_sat ;
 by functionnum ;
 modeleffects predfunction ;
 stderr predstderr ;
run ;

/* 2nd Model: Log linear model with s*p s*c p*c */
proc catmod data=cat2imp1 ;
 by _imputation_ ;
 model c*p*s =_response_  / noparm pred=prob ;
 loglin c p s s*p s*c p*c ;
 ods output predictedprobs=outprobs_m1  ;
run;

/*print out outprobs data set*/
proc print data=outprobs_m1 ;
run ;

/*PROC MIANALYZE to combine predicted prob and SE */
proc sort ; by functionnum _imputation_ ; run ;

proc mianalyze data=outprobs_m1 ;
 by functionnum ;
 modeleffects predfunction ;
 stderr predstderr ;
run ;

/* 3rd Model: Log linear model with s*p s*c */
proc catmod data=cat2imp1 ;
 by _imputation_ ;
 model c*p*s =_response_  / noparm pred=prob ;
 loglin c p s s*p s*c ;
 ods output predictedprobs=outprobs_m2  ;
run;

/*print out outprobs data set*/
```

```
proc print data=outprobs_m2 ;
run ;


/*PROC MIANALYZE to combine predicted prob and SE */
proc sort ; by functionnum _imputation_ ; run ;

proc mianalyze data=outprobs_m2 ;
 by functionnum ;
 modeleffects predfunction ;
 stderr predstderr ;
run ;
</sas>
```

Table 5.4: Proportions from Log-linear Models (SPC), (SP, SC,PC), and (SP, SC), Fitted to Completed Data

		Survival Status	
Clinic	Prenatal Care	Died	Survived
(a) Model: (SPC)			
A	Less	0.0050	0.2578
A	More	0.0076	0.3881
B	Less	0.0259	0.2813
B	More	0.0037	0.0304
(b) Model: (Main Effects plus SP,SC,PC)			
A	Less	0.0047	0.2581
A	More	0.0079	0.3878
B	Less	0.0262	0.2810
B	More	0.0034	0.0016
(c) Model: (Main Effects plus SP,SC)			
A	Less	0.0093	0.3637
A	More	0.0034	0.2823
B	Less	0.0217	0.1755
B	More	0.0079	0.1362

Proportions from each of the three log-linear models are presented in Table 5.4. Model (a) is a saturated model including all interactions terms (S*P*C) while models (b) and (c) include main effects plus selected interactions.

5.6.2   Weighted Least Squares

This weighted least squares example uses data from Stokes, Davis, and Koch, (2001). This data set includes 3 categorical variables measuring cold symptoms

in two discrete periods by gender and urban/rural county status. The data
has been modified such that randomly assigned missing data is present for
the periods with cold symptoms variable. The data set contains 3 variables: 1.
Sex (1=Female and 2=Male), 2. Residence (1=Rural and 2=Urban), and 3.
Periods (0=No cold symptoms in either period, 1=cold symptoms in 1 period
and 2=cold symptoms in 2 periods).

The analytic aim is to use the weighted least squares technique to examine
the mean number of periods with cold symptoms by gender and rural/urban
county status. The analysis is carried out with IMPUTE to handle missing
data imputation, PROC CATMOD to perform weighted least squares modeling using completed data sets, and PROC MIANALYZE to combine results
from PROC CATMOD.

```
<sas name="Weighted Least Squares">
/* Example 3 : Weighted Least Squares Analysis*/
/* Data is from Stokes, Davis, and Koch (2001), missing data

randomly assigned for Periods variable */

%macro it3 (sex, res, period, count) ;

data cat3 ;
do i=1 to &count ;
 sex=&sex ;   residence=&res; period=&period ; count=&count ;
output ;
end ;
drop i count ;
run ;

proc append base=catall3 data=cat3 ;
proc print ;
run ;

%mend ;
%it3(1,1,0, 45) ; %it3(1,1,1, 64) ; %it3(1,1,2,71) ;
%it3(1,2,0, 80) ; %it3(1,2,1, 104) ; %it3(1,2,2,116) ;
%it3(2,1,0, 84) ; %it3(2,1,1, 124) ; %it3(2,1,2,82) ;
%it3(2,2,0, 106) ; %it3(2,2,1,117) ; %it3(2,2,2,87) ;

data cat3_m ;
 set catall3 ;
 if ranuni(87655) <=.10 or ranuni(87655) >=.98 then
 period=. ; else period = period ;
run ;

proc freq data=cat3_m ;
```

```
  tables sex* residence * period / missing list   ;
run ;

proc mi nimpute=0 data=cat3_m ;
run ;
/* impute missing data */
<impute name="Ex3_impute">
title "Categorical Example 3 Imputation for Three Way Table" ;
  datain cat3_m ;
  dataout cat3imp all;
  default categorical;
  interact sex*residence sex*period residence*period ;
  iterations 5;
  multiples 10 ;
  seed 54323 ;
run;
</impute>
```

The above commands prepare a SAS data set from the a tabular summary,
randomly simulate missing data on the "PERIOD" variable, and then impute
missing data using IMPUTE.

```
/*sort data and execute SASMOD with PROC CATMOD */

proc sort data=cat3imp ; by _mult_ ; run ;
proc freq data=cat3imp ;
 by _mult_ ;
 tables sex* residence * period / missing list   ;
run ;

<sasmod name="Ex3_catmod">
datain cat3imp ;
title "Main Effects and Interaction Model" ;
proc catmod ;
 response means ;
 model period =sex residence sex*residence / design ;
run;
</sasmod>

<sasmod name="Ex3_1_catmod">
datain cat3imp ;
title "Main Effects Model" ;
proc catmod ;
 response means ;
 model period =sex residence  / design ;
run;
```

```
</sasmod>
</sas>
```

In the previous set of commands, after preparing the data, the SASMOD command is run twice, first with main effects plus the two way interaction of sex and residence and second, with just main effects because the interaction term in the first model is non-significant. The PROC CATMOD syntax includes the RESPONSE MEANS statement to request means for the response variable, PERIOD. Also, the /DESIGN option produces a design grid in the SAS output (not shown here) but since the SASMOD default output consists of parameter estimates and associated statistics, they are presented in Table 5.5. As previously mentioned, two models are tested using PROC CAT-

Table 5.5: Results from Weighted Least Squares Analysis of Mean Response, Cold Symptoms Data, Main Effects Model

Variable	Estimate	SE	Wald Test	p
Intercept	1.037	0.008	18460.705	0.000
Female	0.071	0.008	85.575	0.000
Rural	0.027	0.008	12.794	0.000

MOD with SASMOD. The first includes main effects plus the interaction of sex*residence but since this interaction is not significant, the model including just main effects is considered the best model.

Based on the parameter estimates in Table 5.5, predicted values for specified types of individuals can be obtained as usual, that is, the expected mean number of periods with cold symptoms for rural females is: 1*1.037 + 1*0.071 + 1*0.027=1.135. For males, the calculation would be 1*1.037 - 1*0.071 + 1*0.027=.993.

5.7   Additional Reading

For more theoretical background on categorical data analysis, refer to Agresti (2012), Bishop, Fienberg and Holland (1975), and Grizzle,Starmer and Koch(1969). For practical guidance on how to analyze categorical variables using SAS see Stokes, Davis, and Koch (2001).

5.8   Exercises

1. Download the **EX_C5_MTF_2014** data set from the book web site. This data set is based on the Monitoring the Future 2014 Grade 12 data and contains a subset of categorical variables for use in these exercises. The goal is to use this data set to perform categorical data analysis using the techniques covered in this chapter.

   (a) Explore the data set and determine how many variables are fully observed and have some missing data. Make sure to do frequency table analyses to evaluate variable distributions and extent of missing data problems.

   (b) Which variables are fully observed and which have some missing data? How much missing data exists on those not fully observed?

   (c) Consider a cross-tabulation of Sex and Ever Smoke Cigarettes in lifetime. Prepare a two way contingency table of these two variables and provide information about observed and missing data for each cell of the table.

   (d) Impute missing data in the exercise data set using a reasonable number of imputations and iterations. Make sure to transfer the case ID variable in the imputation. Use either IVEware or software of your choice.

   (e) Using the completed data sets from (d), run a two way contingency table of Sex*Ever Smoke Cigarettes and obtain a ChiSquare statistic for each multiple.

   (f) Use software or a manual approach (your choice) to calculate a combined Chi-Square statistic. Describe the relationship between sex and ever smoke cigarettes and answer if sex and ever smoke cigarettes are statistically independent. Provide support for your conclusion.

2. Create a subset of data from the MTF data used in Exercise 1 and retain three variables: v2150, v2101d, and v2150. The goal of this exercise is to perform a three way contingency table analysis of sex, race, and ever smoke cigarettes, using log-linear models with multiply imputed data.

   (a) Explore the missing data problem with software of your choice and determine how much missing data exists and how many imputations are appropriate.

   (b) Impute missing data using IMPUTE or other software and state how many imputations/iterations are used, the SEED value used, and which interactions you included in the imputation models. What is the value of adding interactions to the imputation?

(c) Prepare a 3 way contingency table using the completed data sets and make sure to use DESCRIBE or similar approach to obtain combined cell proportions. Save these results for a comparison to output from log-linear models to come.

(d) Use the completed data sets and run a log-linear model (fully saturated) to explore relationships between the 3 categorical variables of choice. Use SASMOD with PROC CATMOD or PROC CATMOD and PROC MIANALYZE (SAS Users) or similar tools in your software of choice if not using SAS. Obtain combined expected cell proportions from the fully saturated log-linear model and compare to the results in part (c). Do the proportions match and if so, is this expected?

(e) Repeat part (d) but rather than a fully saturated model, use a model with all 2 way interactions but not a 3 way interaction. How do the cell proportions compare to those from part (d)? How would you explain the differences/similarities?

3. Again, use data from Exercise 1 and create another subset of data retaining sex (V2150), high school type (HSTYPE), and number of times drink alcohol in lifetime (NUMALC). The goal of this exercise is to perform a weighted least squares analysis using categorical variables.

The variable NUMALC is coded 0=no drinks in lifetime, 1=1-9 drinks in lifetime, or 2=10+ drinks in lifetime and the interest is in how sex and Vocational/Other high school affects the mean number of drinks (categorical) had by grade 12. For the analysis, create a new variable called VOC coded 1 if HSTYPE=3, '.' if HSTYPE=., and 2 if HSTYPE =1 or 2. Note that VOC represents both Vocational and Other types of high school. Retain just V2150, VOC, and NUMALC.

(a) Prepare the needed subset of data and explore the missing data problem. How would you describe the missing data pattern?

(b) Impute missing data using your chosen software. Using the imputed data sets, prepare a three way contingency table with DESCRIBE or a similar tool capable of correct combining. Make sure that all missing data is now imputed.

(c) Use PROC CATMOD with SASMOD (or equivalent) to run a weighted least squares model with main effects and the interaction of sex*vocational HS, similar to Example 3 of this chapter. Make sure to request a mean response and obtain parameter estimates using correct combining rules. Why is it acceptable to use a linear modeling technique with a categorical outcome and what assumptions are made in doing so? Is the interaction term significant? If not, rerun the model with just main effects of sex and Vocational/Other HS.

(d) Prepare a table of regression estimates similar to Table 5.5 and provide estimates of number of drinks during lifetime for a few of the sex and Vocational/Other HS profiles. How would you calculate and interpret the predicted value for males in Vocational/Other high school?

6

Survival Analysis

6.1 Introduction

The survival analysis involves regression models where the outcome is time to an event such as death, free of disease, divorce, marriage etc. In an industrial setting, the time-to-event may be failure of one or more devices or components. The distinguishing feature of this type of analysis is that the time-to-event may not occur during the observation or study period. In this case, the observation is right censored.

The time-to-event outcomes are generally skewed due to more events occurring in the early time period and with a long right tail. Censored observations only provide a lower bound for the actual time-to-event yet to occur. For these reasons, typically the mean function is not modeled but the whole distribution is modeled through a feature related to the distribution function called the hazard function. Suppose that T is the time-to-event and C is the censoring time (end of the study, subject dropped out and no further information is available etc.). Let $Y = Min(T, C)$ be the actual outcome observed. That is, $Y = T$ if the event were to occur during the study period (or known to have occurred even if beyond the study period), otherwise $Y = C$, the last known time without having the event occurrence. For now, assume that everybody is subject to the event occurring (death, for example). The goal is to study the relationship between T and X given (Y, X). Note that this situation can be treated as a missing data problem where T is observed for uncensored subjects and is missing among censored subjects although the missing value is known to be greater than C.

Let the distribution function of T be $F(t) = Pr(T \leq t)$. The survival function is $S(t) = 1 - F(t) = Pr(T > t)$ and the density function is $f(t) = dF(t)/dt$ or $F(t) = \int_0^t f(u)du$. The hazard function measures the likelihood of the event occurring in the "immediate future" given that it has not occurred up until now. Let t be the current time and the immediate future is a small time interval $(t, t + dt)$. The conditional probability of the event occurring during this interval given that the event has not occurred until t is

$$\frac{Pr(t \leq T \leq t + dt)}{Pr(T > t)} = \frac{F(t + dt) - F(t)}{S(t)} = \frac{f(t)dt}{S(t)} = \lambda(t)dt$$

for an arbitrarily small dt where $\lambda(t) = f(t)/S(t)$ is called the hazard function.

The goal of the survival analysis is to study the relationship between $\lambda(t)$ and the covariates X.

A popular model for the hazard function is the proportional hazards or Cox model. Let $\lambda_o(t)$ be the "baseline" hazard function which is modified by the covariates through the relationship,

$$\lambda(t|x) = \lambda_o(t) \exp(x^T \theta).$$

Suppose that $x = 1$ or 0. The $\lambda_o(t)$ is the hazard function for $x = 0$ and it is proportionately high or low for $x = 1$ across the entire time period by the quantity $\exp(\theta)$, called hazard ratio or relative risk. Cox developed a clever conditioning argument which compares subjects having an event to the risk set of all those exposed to the possibility of having the event at that moment. This argument developed a partial likelihood that did not require knowledge of $\lambda_o(t)$. Proportionality of the hazard is a strong assumption but the benefit of not modeling $\lambda_o(t)$ results in considerable robustness.

In engineering and other fields, parametric models are used for $f(t)$ as a function of covariates. These can be handled using the generalized linear model framework. The analysis will impute simultaneously the censored outcomes and missing covariates in the multiple imputation analysis. Another model often employed is the Tobit model which involves left or right truncation. Examples of both Cox and Tobit models (with right truncation) are presented later in this chapter.

6.2　Multiple Imputation Analysis

The foregoing description of the survival analysis makes it clear that the outcome is not fully observed with censored observations. However, partial information is available in the form of a lower bound for the actual failure time or time-to-event. The censoring time is that lower bound. Thus, the outcome itself is missing with partial information. In addition, covariates may also be missing.

How should the imputation of missing values then proceed? There are two possible options:

1. Impute jointly the outcome and missing covariates treating the censoring time as a lower bound for the missing outcome variable (censored cases).

2. Treat censoring/failure time and censoring indicators as covariates in the imputation of covariates. That is, do not impute the outcome variable.

Under approach (1), a joint model for $Pr(T, X)$ is used in the imputation process where the censoring time C is incorporated when imputing the missing values in T. Whereas approach (2) uses the conditional model $Pr(X|Y, D)$, where D is a censoring indicator variable, in the imputation process. Furthermore, under approach (1), we have the option of performing the analysis using the multiply imputed T or to ignore the imputed values in T but use (Y, D) and multiply imputed X in the analysis. The choice between the two approaches depends on whether one considers (Y, D) as the fully observed outcome variable or as a missing outcome variable. In the next section, a particular data set will be analyzed both ways to assess the differences between the two approaches.

6.2.1 Proportional Hazards Model

The Proportional Hazards model is demonstrated using data from the Primary Biliary Cirrhosis data set. As previously described, the data are from a clinical trial consisting of 418 patients where 312 agreed to be randomized to either a placebo or to the drug D-penicillamine and another 106 patients agreed to be followed to death.

Two methods are used. The first treats missing data on survival time and model covariates such as Ascites, Serum Cholesterol, etc. as an imputation task and imputes missing data on the covariates as well as survival time for censored respondents. In the imputation of survival time for censored cases, the time to censor is used as a lower bound along with an upper bound of 25 to prevent imputation of very large survival times.

The second method uses censor time and an indicator of being censored as predictors in the imputation model but does not impute missing data on survival time. Using the completed data sets, Cox models are executed using the *IVEware* REGRESS command.

These examples demonstrate use of *IVEware* with SPSS. Detailed code snippets are presented and explained within each section. Broadly speaking, the first two examples perform multiple imputation of missing data using IMPUTE using methods 1 and 2 described above. The 25 completed data sets are then analyzed using REGRESS with the LINK PHREG and CENSOR CENSORED(1) statements to declare the type of regression desired and identify censored cases with CENSORED=1 syntax. The final example demonstrates imputation and execution of a Tobit model with right truncation.

6.2.1.1 Outcome Imputed (Method 1)

```
<spss name="Survival Analysis Cox Model">
/* run multiple imputation using censored/failure
 time as outcome imputed*/
<impute name="impute">
title "Survival Analysis, Outcome Imputed" ;
datain  pbcimpute_28may2016;
```

```
dataout pbcdata_out all;
default continuous ;
transfer id imp_surv ;
categorical sex status censored stage edema ascites
 hepato spiders drug;
bounds lsurv(>= lower, <= 25);
iterations 10 ;
multiples 25 ;
seed 987 ;

run ;
</impute>

<regress name="phreg">
title  "Cox Model with Log Survival as Outcome Imputed" ;
datain pbcdata_out ;
link phreg ;
categorical drug ;
dependent lsurv ;

predictor drug ageyrs sex ascites hepato spiders edema
 lbili lchol albumin lcopper lalk_phos
lsgot ltrig  platelet protime stage ;
run ;
</regress>
```

The preceding syntax imputes missing data on the outcome and covariates using IMPUTE and then analyzes the completed data sets using REGRESS with the PHREG link along with other regression options previously discussed.

Table 6.1 presents selected output based on imputation of outcomes and covariates, using the Proportional Hazards model.

6.2.1.2   Outcome Not Imputed (Method 2)

```
/*multiple imputation with censor/failure time and censor
 indicator as predictors in model*/
<impute name="impute_m2">
title "Survival Analysis, Method 2" ;
datain  pbcimpute_12aug2016;
dataout pbcdata_out_m2 all;
default continuous ;
transfer id ;
categorical sex status censored stage edema ascites
 hepato spiders drug;
iterations 10 ;
multiples 25 ;
```

Table 6.1: Results from Regression Analysis of PBC Imputed Data, Outcome Imputed

Variable	Hazard Ratio	Confidence Interval
Treatment	0.902	(0.830, 0.981)
Placebo	0.951	(0.871, 1.039)
Age in Years	1.034	(1.031, 1.038)
Gender	1.100	(0.990, 1.223)
Ascites	1.469	(1.325, 1.629)
Hepatomegaly	1.108	(1.023, 1.200)
Spiders	0.993	(0.922, 1.069)
Edema	2.241	(1.993, 2.520)
Log Bilirubin	1.883	(1.785, 1.986)
Log Cholesterol	1.051	(0.958, 1.152)
Albumin	0.631	(0.577, 0.690)
Log Copper	1.473	(1.398, 1.551)
Log Alkaline Phosphate	0.935	(0.893, 0.979)
Log Serum glutamic-oxaloacetic	1.489	(1.357 1.633)
Log Triglicerides	0.875	(0.808, 0.947)
Platelet Count	1.000	(0.999, 1.000)
Prothrombin time (seconds)	1.206	(1.171,1.242)
Histologic Stage	1.385	(1.316, 1.457)

```
seed 987 ;
run ;
</impute>

<regress name="phreg_m2">
title  "Cox Model with Log Survival as Outcome Not Imputed" ;
datain pbcdata_out_m2 ;
link phreg ;
categorical drug ;
dependent lsurv ;
censor censored(1) ;
predictor drug ageyrs sex ascites hepato spiders edema
 lbili lchol albumin lcopper lalk_phos lsgot
ltrig  platelet protime stage ;
run ;
</regress>
```

The syntax above imputes missing data using method 2 where censor/failure time and the censor indicator are used as model covariates but not imputed. This step is again followed by use of a Cox model from the REGRESS command. Results from the second Cox model are presented in Table 6.2.

Table 6.2: Results from Regression Analysis of PBC Imputed Data, Outcomes Not Imputed

Variable	Hazard Ratio	Confidence Interval
Treatment	0.884	(0.812, 0.962)
Placebo	0.932	(0.853, 1.018)
Age in Years	1.035	(1.031. 1.039)
Gender	1.105	(0.994,1.228)
Ascites	1.549	(1.394 ,1.200)
Spiders	0.965	(0.895, 1.040)
Edema	2.345	(2.084, 2.637)
Log Bilirubin	1.855	(1.757,1.959)
Log Cholesterol	1.131	(1.031,1.240)
Albumin	0.636	(0.582 ,0.696)
Log Copper	1.479	(1.403, 1.559)
Log Alkaline Phosphate	0.918	(0.876, 0.962)
Log Serum glutamic-oxaloacetic	1.522	(1.385, 1.672)
Log Triglicerides	0.847	(0.783, 0.916)
Platelet Count	1.000	(1.000, 1.001)
Prothrombin time (seconds)	1.210	(1.175, 1.246)
Histologic Stage	1.387	(1.319, 1.459)

Results from Tables 6.1 and 6.2 are similar and both indicate that there is not much difference in risk between the Treatment group or Placebo group participants, as compared to the Observed group. Significant high risk factors are Age, presence of Edema, Ascites, elevated Serum Bilirubin, Cholesterol, Copper, Serum GO, Prothrombin time, and Histologic Stage.

6.2.2 Tobit Model

The Tobit model example uses a hypothetical data file, **tobit_miss**, containing 200 observations and simulated missing data on select variables, solely for demonstration purposes. The simulated data is based on a data set obtained from www.ats.ucla.edu/stat/sas/. The goal is to regress aptitude on reading and math scores with type of school program attended.

The variable APT represents academic aptitude while reading and math test scores are contained in the variables READ and MATH, respectively. The variable PROG is the type of educational program attended and is coded as academic (PROG = 1), general (PROG = 2), or vocational (PROG = 3). Missing data exists on two variables, READ and MATH with fully observed data for PROG and APT. The theoretical range for APT is 200-800, however, there are no observed scores of 200 while some scores are truncated at 800. A score of 800 is considered upper truncation but since there are no values of 200, lower truncation is not seen in this data.

The following syntax imputes missing data with the IMPUTE command and uses REGRESS with a TOBIT link to predict academic aptitude by reading and math scores and program type (with Vocational program the reference group). Note the use of the closing tag (</spss>) to end the SPSS session for the entire set of examples executed in this chapter.

```
/* Tobit Model using Right Truncated Academic
Aptitude Data with Missing Data on Covariates */
<impute name="impute_tobit" >
title "Impute Tobit Data" ;
datain  tobit_miss ;
dataout tobit_imputed all;
default continuous ;
transfer id ;
categorical prog ;
iterations 5;
multiples 10;
seed 55;
run ;
</impute>


<regress name="Tobit Model with Imputed Missing">
title  "Imputed Missing Data Tobit Model" ;
```

```
datain tobit_imputed ;
link tobit ;
categorical prog ;
dependent apt;
predictor read math prog;
run ;
</regress>
</spss>
```

Table 6.3: Results from Tobit Model of Academic Aptitude

Variable	Parameter Estimate	Standard Error	Wald Test	p Value
Read	2.907	0.186	242.808	0.000
Math	4.951	0.211	550.543	0.000
Academic Program	50.551	4.136	149.350	0.000
General Program	38.767	3.874	100.128	0.000
Sigma	62.912	0.995	4000.000	0.000

The results presented in Table 6.3 indicate that a one unit change in reading scores results in about 3 additional points on the aptitude variable and about 5 additional points due to a one unit change in Math score. For program type, compared to attending a Vocational program, being in an Academic program results in a 50 point increase in the aptitude score while being in a General program results in a 39 point increase in the outcome. Sigma can be used as a comparison to the standard deviation of academic aptitude which was 99.21, a substantial reduction. For more on this statistic, see Tobin (1958).

6.3   Additional Reading

For more information about Tobit models, see Amemiya (1984) and Tobin (1958). Good references for survival analysis include Fleming and Harrington (2005), Hosmer,Lemeshow and May (2008), and Kalbfleisch and Prentice (2002). Lavori, Dawson and Shera (1995), van Buuren, Boshuizen and Knook (1999), White and Royston (2009) deal with missing data in survival analysis.

6.4 Exercises

1. Download the data set named **EX_NCSR_SURVIVAL** from the book web site and examine the contents, variable type, and missing data present in this data set. The analysis goal is to predict incidence and onset of Major Depressive Disorder using data from Part 2 of the NCS-R (n=5,692). The data set includes a number of variables with some missing data. The outcome variable is age of onset for those that had a diagnosis of MDE or age at interview for those without a MDE diagnosis (considered right-censored).

   (a) Based on exploratory analysis, which variables have missing data? What is the highest percentage of missing data? Which variables are categorical, continuous, or any other type? How many imputations would you recommend given the missing data rates? Describe what the AGEEVENT and CENSOR variables represent.

   (b) Impute missing data using the censor indicator (CENSOR) and the failure/censor time (AGEEVENT) variables as predictors in the imputation. Use M=5 with 5 iterations and a SEED=123 value. Request imputation diagnostics for the PEA76 variable.

   (c) Examine the imputation results and diagnostic plots with the goal of identifying any issues with the imputation. Do you see any issues that require investigation?

   (d) Perform survival analysis using the 5 imputed data sets from part b. and request a PH model from the REGRESS command. The model of interest is incidence and onset of Major Depressive Episode (AGEEVENT) predicted by gender (SEX), education in 4 categories (EDUCAT), PEA76, PEA79, PEA80, and PEA82. Make sure to use the censor indicator variable (CENSOR) set to 1 to identify the censored cases, specify LINK PHREG to request a PH model, incorporate the complex sample design features through use of the weight, stratum, and cluster variables and declare variable type prior to running the model.

   (e) Based on the results from part d., write a short paragraph describing how missing data was handled, and how REGRESS handles combining MI results and also accounts for complex sample design features and interpretation of the results.

2. Download the **EX_WHAS500_SURVIVAL** data set (SAS format) from the book web site. For this exercise, the data set includes randomly simulated missing data on a few covariates. The original

data is based on the Worcester Heart Attack Study WHAS500 Data, directed by Dr. Robert J. Goldberg of the Department of Cardiology at the University of Massachusetts Medical School. The data set was obtained from Hosmer, D.W. and Lemeshow, S. and May, S. (2008). The main goal of this study is to describe factors associated with trends over time in the incidence and survival rates following hospital admission for acute myocardial infarction (MI). Data have been collected during thirteen 1-year periods beginning in 1975 and extending through 2001 on all MI patients admitted to hospitals in the Worcester, Massachusetts Standard Metropolitan Statistical Area. (Hosmer and Lemeshow, 2009).

(a) The aim of this exercise is to use a Cox model to predict incidence and survival rates after hospital admission due to acute Myocardial Infarction. As usual, prepare for imputation and analysis through data exploration including identifying missing data patterns, extent of missing data, and variable type. Which variables have missing data and what are the missing data rates? What do the values on the LENFOL variable mean for a censored case? What are two methods of multiple imputation that could be used to handle censored outcome cases?

(b) Impute missing data using the IMPUTE command. For this imputation, use the failure/censor time and censor indicator variables in the imputation model as predictors. Create M=10 imputed data sets, and include a SEED value to ensure future replication of results. Make sure perform imputation diagnostics for each imputed variable using the DIAGNOSE command. Are there any issues with the imputation and if so, address these prior to regression analysis of imputed data sets.

(c) Use the 10 imputed data sets as input to the REGRESS command and perform survival analysis using a Cox model. The model of interest is: LENFOL = BMI HR AFB. Be sure to use the LINK PHREG statement and the correct specification of the dependent variable in the syntax.

(d) Repeat parts b. and c. but treat censored survival times as missing and impute as an outcome variable. Make sure to create a new outcome variable set to missing for censored cases, set the lower bound to the censor followup time and the upper bound to the maximum number of days in the observed data. Also impute missing data on covariates as well.

(e) Based on the results from parts c. and d., prepare a table of results from the two approaches. Discuss and interpret the findings and be sure to cover how multiple imputation improves the analysis and how MI variability is included in the analysis of complete data sets.

3. Download the **EX_TRUNCREG_MISS** data set from the book web site. The data set is derived from a hypothetical study of students in the GATE (gifted and talented education) program. The goal is to model achievement as a function of language skills and the type of program in which the student is enrolled. Since all students are required to have a minimum achievement score of 40 to enter the GATE program, the sample is truncated at an achievement score of 40. An additional complication is there is missing data on some variables in the data set.

   (a) Perform exploratory data analysis on each variable to determine variable type and contents, missing data rates, and missing data patterns. How many variables have missing data and what is the pattern? How many imputations would you recommend?

   (b) Impute missing data and include imputation diagnostics for each variable requiring imputation. Evaluate the diagnostic plots and determine if any imputation issues need to be addressed before regression modeling.

   (c) Perform Tobit regression predicting achievement by language scores and type of program. Run the model once using complete cases only and repeat using imputed data sets produced in part b. Write a brief comparative summary of the results and describe how inherent lower truncation was dealt with, how multiple imputation improves the analysis and what the Tobit model results suggest about the relationships between achievement and the explanatory variables.

7

Structural Equation Models

7.1 Introduction

Structural equation modeling is a technique used to assess whether the correlation matrix of p variables can be explained through a set of pre-specified linear regression relationships. These relationships are typically expressed through a collection of Directed Acyclic Graph (DAG) diagrams. The graph may be connected either through a directional ("causal") or bi-directional ("association") relationship and also may involve latent variables connected to observed variables. Figure 7.1 shows a collection of typical symbols used in expressing the relationship through a DAG. The letters (Y) in the square boxes are used for observed variables (which are expressed as Z-scores), the letters in the circle (F) are latent variables which are constructs related to observed variables and letters E without the boxes or squares represents residuals or unexplained variation. The one or two sided arrows are used to indicate dependence between the variables in the boxes.

Figure 7.1: Schematics used in specifying structural equation models

Figure 7.2 provides a schematic with 6 observed variables $Y_1, Y_2, \ldots, Y_6$, two latent variables F_1 and F_2 and 6 residuals $E_1, E_2, \ldots, E_6$.

The first three observed variables are related to the latent construct F_1 through the following regression relationships,

$$
\begin{aligned}
Y_1 &= \beta_{11} F_1 + E_1 \\
Y_2 &= \beta_{21} F_1 + E_2 \\
Y_3 &= \beta_{31} F_1 + E_3
\end{aligned}
$$

Figure 7.2: An example of a structural equation model specification

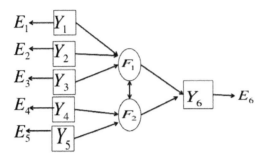

Similarly, the variables Y_4 and Y_5 are related to the latent variable F_2 through the following regression relationships,

$$Y_4 = \beta_{42}F_2 + E_4$$
$$Y_5 = \beta_{52}F_2 + E_5$$

Finally, the last relationship is between Y_6 and the two latent variables F_1 and F_2,

$$Y_6 = \beta_{63}F_1 + \beta_{64}F_2 + E_6.$$

The two latent variables are assumed to have mean 0, variance 1 and the bi-directional arrow indicates that they are correlated with a correlation coefficient, τ_{12}. The six residuals, $(E_1, E_2, \ldots, E_6)$, are assumed to be uncorrelated with mean 0 and variances $\sigma_i^2, i = 1, 2, \ldots, 6$. The seven regression coefficients $\beta = (\beta_{11}, \beta_{21}, \beta_{31}, \beta_{42}, \beta_{52}, \beta_{63}, \beta_{64})$ are called path coefficients. The regression models given above imply a structure of the covariance matrix in terms of the residual variances, the correlation coefficient between the latent variables, and the path coefficients. These parameters can be estimated using, for example, the maximum likelihood approach.

Missing values in the observed variables may be multiply imputed using, perhaps, additional covariates that may be predictive of these observed variables. Data analysis can be performed on each completed data set and multiply imputed estimated path coefficients and their standard errors can be obtained using the combining rules discussed in Chapter 1. Note that the imputation should be carried out without assuming any structure on the correlation matrix (arbitrary covariance matrix). Otherwise, the completed data analysis will be biased towards the assumed or particular structure.

7.2 Example

This example uses individual level data from the National Merit Twin Study (n=1,678) and includes scores from National Merit tests and individual/ family characteristics such as parent education, family income, type of twin, and gender. See Appendix A (or Loehlin and Nichols(1976)) for more details about the data set. The goal is to perform structural equation modeling to examine relationships between "manifest" variables representing Mother's and Father's education, Family Income, and English, Social Science, Vocabulary, Math and Natural Science test scores. Three latent variables are included in the analysis: Family Background, Verbal Ability and Quantitative Ability. See Figure 7.3 for a DAG schematic representing these relationships.

Figure 7.3: Schematic for structural equation model using National Merit Twin Study data

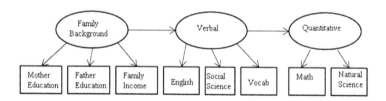

Three approaches are contrasted in this example: (1) Complete case analysis using SAS PROC CALIS, (2) Full Information Maximum Likelihood (FIML) also using PROC CALIS, and (3) Multiple Imputation (MI) with IMPUTE and SASMOD (for use with SAS only). Non-SAS users may impute missing data using the IMPUTE command and analyze output data sets in a software of choice, assuming it can correctly combine data sets from multiple imputation.

The code below first examines the data with a focus on descriptives and the pattern of missing data. Next, PROC CALIS is used for both complete case and FIML SEM analyses. Use of the "path" SAS syntax for defining the relationships is demonstrated though a number of other syntactical options are also available in PROC CALIS. Missing data is imputed using IMPUTE with M=5 and a SEED value. The imputed data set is then used with

SASMOD/PROC CALIS to perform SEM within the structure of the SAS-MOD command. The advantage of using SASMOD with PROC CALIS is that proper combining of MI results is done automatically. The code is presented in full prior to presentation and discussion of results.

```
<sas name="SEM_Example">
libname sem 'P:\ive_and_MI_Applications_Book\Chapter8SEM';

/*check for missing data */
proc means data=sem.merit_14dec2016 n nmiss mean min max ;
run ;

/* PROC CALIS without FIML, CC analysis */
proc calis data=sem.merit_14dec2016 ;
path
 moed faed faminc <--- familybackground = 1,
 english socsci vocab <--- verbal = 1,
 math natsci <--- quantitative = 1,
 familybackground ---> verbal,
 verbal ---> quantitative;
run ;



/* PROC CALIS with FIML*/
proc calis data=sem.merit_14dec2016 method=fiml ;
path
 moed faed faminc <--- familybackground = 1,
 english socsci vocab <--- verbal = 1,
 math natsci <--- quantitative = 1,
 familybackground ---> verbal,
 verbal ---> quantitative;
run ;

/* impute missing data here before proceeding to SEM */
<impute name="SEM_Impute">
title "Impute Missing Data Prior to SEM Analysis" ;
datain  sem.merit_14dec2016 ;
dataout nmtdata_out all;
default continuous ;
transfer pairnum  ;
categorical sex zygosity moed faed faminc;
iterations 5 ;
multiples 5 ;
seed 305 ;
run ;
</impute>
```

```
/* run with PROC CALIS and SASMOD to compare with
Complete Case and FIML */
<sasmod name="SEM_SASMOD">
title  "SASMOD with PROC CALIS for SEM Model" ;
datain nmtdata_out ;
by _mult_ ;
proc calis ;
path
 moed faed faminc <--- familybackground = 1,
 english socsci vocab <--- verbal = 1,
 math natsci <--- quantitative = 1,
 familybackground ---> verbal,
 verbal ---> quantitative;
run ;
</sasmod>
</sas>
```

7.3   Multiple Imputation Analysis

Due to missing data on Mother's education (missing for 38 observations), Father's education (missing on 48 observations), and Family Income (missing on 124 observations), multiple imputation is executed using the IMPUTE command and five complete data sets generated for subsequent analysis. Data analysis is carried out as described above, complete case analysis (missing data cases omitted), use of FIML, and analysis of multiply imputed data sets with SASMOD. The tables below summarize the results. Default output from the SASMOD command includes standardized regression estimates and standard errors, Wald tests, and p values but many other outputs are available from PROC CALIS. See full output from PROC CALIS runs for CCA and FIML for details.

Results from Tables 7.1-7.3 indicate very few differences among the 3 methods. The default results from PROC CALIS (Tables 7.1 and 7.2) include a t test while results from SASMOD/PROC CALIS present a Wald test, that is the t statistic squared. Overall, each parameter is significant at the 0.05 level and the unstandardized estimates differ in the 2nd or 3rd decimal.

Table 7.1: Results from Complete Case Analysis of National Merit Twin Study, n=1,536

Variable	Unstandardized Estimate	t	$\Pr > \|t\|$
Mother Education	1.000	−	
Father Education	1.636	19.832	<.0001
Family Income	1.212	19.502	<.0001
English	1.000	−	
Social Science	1.229	34.920	<.0001
Vocabulary	1.170	34.293	<.0001
Math	1.000	−	
Natural Science	0.975	31.448	<.0001
Family Background	1.636	11.441	<.0001
Verbal	1.221	26.330	<.0001

Table 7.2: Results from FIML Analysis of National Merit Twin Study, n=1,678.

Variable	Unstandardized Estimate	t	$\Pr > \|t\|$
Mother Education	1.000	−	
Father Education	1.612	20.551	<.0001
Family Income	1.203	19.919	<.0001
English	1.000	−	
Social Science	1.223	35.578	<.0001
Vocabulary	1.179	36.034	<.0001
Math	1.000	−	
Natural Science	0.984	32.792	<.0001
Family Background	1.592	11.609	<.0001
Verbal	1.218	27.644	<.0001

Table 7.3: Results from MI Analysis of National Merit Twin Study, n=1,678.

Variable	Unstandardized Estimate	Wald	$\Pr > \|t\|$
Mother Education	1.000	−	
Father Education	1.607	426.649	<.0001
Family Income	1.196	404.750	<.0001
English	1.000	−	
Social Science	1.223	1320.388	<.0001
Vocabulary	1.180	1283.387	<.0001
Math	1.000	−	
Natural Science	0.984	1075.014	<.0001
Family Background	1.571	136.041	<.0001
Verbal	1.218	759.994	<.0001

7.4 Additional Reading

Key references for Structural Equation Modeling include Kline(1998), Schumacker, and Lomax(1996), and Bollen(1989). Also see Zhang and Yung (2011) for SAS conference publications, tutorials, and related examples. Enders (2006, 2010) provide an excellent overview of missing data methods for Structural Equation modeling. See also, Allison (2003), Bodner (2008) and Schminkey, von Oertzen and Bullock (2016).

7.5 Exercises

1. Download the data set called "**Ex_sem**" from the book web site and translate for use with your software of choice. The data was obtained from SAS technical support (support.sas.com) and was designed for use in a number of papers by Zhang and Yung (2011). The simulated data closely mimics actual data described in detail in Marjoribanks (1974). The point of the analysis is to use SEM to model mental ability using 12 manifest variables and 4 latent variables. A path diagram below represents the model relationships.

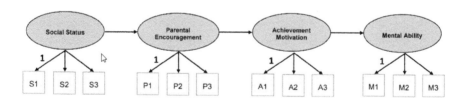

The diagram details relationships with measured variables represented by squares and latent variables represented by ovals. The data set contains 200 observations and was generated from a multivariate normal distribution for the 12 observed variables. However, 50% of the observations have random missing values, (Zhang and Yung, 2011). This exercise uses both MI and FIML with SASMOD to address missing data problems and perform SEM with imputed data. Note that PROC CALIS allows built-in use of FIML and the *IVEware* SASMOD command permits use of PROC CALIS within the Jackknife Repeated Replication structure of *IVEware*.

(a) Explore the data set via descriptive analysis, examination of missing data patterns, variable type, and extent of missing data.

What are the variable types in the data set? Which variables have the highest and lowest amount of missing data?

(b) Impute missing data using IMPUTE with M=15/5 iterations, bounds for each variable, and obtain imputation diagnostic plots for 3 of the 12 variables in the data set. Do any of the imputation diagnostic plot indicate problems with the imputations? If so, how might you address this issue?

(c) **SAS Users:** use Iveware/SASMOD and SAS/PROC CALIS to perform SEM using the imputed data from part b., while following the path diagram above as a guide. PROC CALIS allows use of a number of code methods to set up the analysis. Any code method is acceptable assuming it expresses the same relationships.

Non-SAS users: with your chosen software and the imputed data sets from part b., perform SEM as detailed above. Make sure to follow the path diagram above and also make sure your software is capable of combining results from multiply imputed data sets used in the analysis.

(d) Based on the multiple imputation results from part c., prepare a summary table of path parameters, standard errors and either t or Wald tests with p values. Explain how the variability introduced by the imputation is accounted for and also describe the SEM model in broad terms.

(e) **SAS Users:** repeat the analysis in part d. but rather than using MI to address missing data issues, use PROC CALIS with FIML. Add the parameter estimates, standard errors and significance test information to the table prepared in part d. and label the results "FIML with CALIS". Do the MI v. FIML results differ substantially? If so, describe the key differences or if not, state how they are similar. How do the two approaches compare in terms of general method?

2. **Project**. Generate several complete data sets with 6 variables and DAG given in Figure 7.2 (choose any convenient values for the residual variances (for example, $\sigma_i^2 = 1, i = 1, 2, \ldots, 6$)and path coefficients (for example, all β's=1). Assume that Y_6 has no missing values and all other variables have missing values with the mechanism depending upon Y_6. Use 5 logistic models to generate varying amount of missing data in $Y_1, Y_2, \ldots, Y_5$. Generate five additional standard normal variables, Z_1, Z_2, Z_3, Z_4, Z_5, where Z_i is correlated with Y_i with the correlation coefficient, $\rho_i, i = 1, 2, \ldots, 5$.

(a) One each complete data set (before deleting the values) fit the SEM model, store the path coefficients, their standard errors and 95% confidence intervals.

(b) On each data set with missing values, apply complete case analysis, store the estimated path coefficients, their standard errors and 95% confidence intervals.

(c) Apply FIML approach on each data set with missing values and store the same results.

(d) Multiply impute the missing values in each data set (ignoring the auxiliary variables Z's) and perform multiple imputation analysis. As before store the results.

(e) Repeat (d), except now include the Z's in the imputation model.

(f) Analyze the results from the above simulation study to compare the bias, variance and coverage properties of various approaches. In particular, explain the benefits, or lack thereof, in using the auxiliary variables. To assess the full impact, you may change the amount of missing values and the correlations between Y and Z.

(See Chapter 11 on how to run simulations using *IVEware*.)

8

Longitudinal Data Analysis

8.1 Introduction

A longitudinal study involves collecting some variables repeatedly over time (waves or periods) on a sample of subjects. Such investigations may use different designs such as observational studies like surveys, measurement of biomarkers, clinical measures, health conditions etc. Examples are the panel surveys, prospective cohort studies etc. Experimental longitudinal studies are also common where interventions are mounted in between the time periods. Examples include cross over designs, adaptive treatment regimens etc. New variables may also be added to the data collection at any wave. These designs allow for investigating trends and efficient estimation of treatment effects with a better control for confounding through within-subject comparisons. However, a big problem is missing data due to dropouts. In addition, there may be item-missing data due to refusal to answer some questions or inability in obtaining certain measures.

Analysis of incomplete data in a longitudinal setting can be a real challenge. Consider a simple randomized study with a single baseline covariate X, a treatment indicator $T = 1$ (Treatment) or $T = 0$ (Control) and p waves or periods of outcome measurements, $Y: Y_1, Y_2, \ldots, Y_p$. The subjects may either drop out after a particular wave permanently or intermittently miss some waves.

Suppose the goal is to estimate the parameter,

$$\theta = E(Y_p - Y_1 | T = 1, X) - E(Y_p - Y_1 | T = 0, X)$$

$$= [E(Y_p | T = 1, X) - E(Y_p | T = 0, X)] - [E(Y_1 | T = 1, X) - E(Y_1 | T = 0, X)].$$

Consider several scenarios:

1. Missing Values: (A) No missing values in X and all the missing values are in the outcome variables $Y_1, Y_2, \ldots, Y_p$;or (B) Missing values are in both X and Y's

2. Type of analysis: (A) Intent to treat (as randomized) or (B) as treated analysis

3. Distribution of Outcome variables: (A) Normal or (B) Non-normal

For the situation described by the scenario (1A, 2A, 3A), the maximum likelihood approach (such PROC MIXED in SAS) could be used to infer about the parameter, θ. For a binary outcome variable, technically the scenario (1A, 2A, 3B), a log-linear type model could be formulated and the maximum likelihood estimates could be constructed, though specialized computer code will have to be developed to implement this approach. It is very difficult, if not impossible to develop maximum likelihood estimates for more complex situations such as time-varying covariates with missing values. The multiple imputation approach might be more attractive across all the scenarios.

First, consider the multiple imputation approach for (1A, 2A, 3A) situation. Create a data file with one line per subject with $p + 2$ variables: X, T and $Y_j, j = 1, 2, \ldots, p$. Multiply impute the missing values in the data set using the SRMI approach. For the analysis, create the difference variable $Z = Y_p - Y_1$ and then regress Z on T and X. The regression coefficient for T is the treatment effect θ given above. This is the classical MAR analysis where the drop out and completers are treated as exchangeable within the assigned treatment group and is equivalent to the maximum likelihood analysis. This strategy, "imputation as randomized", can be adopted for all other scenarios involving 2A.

Now consider the modification where dropouts are assigned to the control condition at the time of drop out for imputation purposes. This strategy can be called "imputation as control". Create a data file with one line per subject with $(2p + 1)$ variables: X, $(Y_j, T_j), j = 1, 2, \ldots, p$ where $T_j = 1$, for those subjects who did not drop out and for those who dropped out at wave j, $T_j = 0$ and Y_j is set to missing value. The same strategy works for all other scenarios involving "drop outs considered as control" analysis. The missing values in X can be imputed along with the Y values. The analysis strategy is the same as given above (regressing Z on T and X). This analysis, however, estimates the treatment effect between the two randomized groups where the dropout outcome pattern will be more like controls rather than the treated (as assumed in the "imputation as randomized" analysis). The two analyses can provide deeper insights into the effect of treatment on the outcome.

There are many other assumptions that can be made while assessing the effect of the treatment on the outcome. For example, the dropout being stable post dropout is an extension of the Last Observation Carried Forward (LOCF) method. Creation of completed data sets under various assumptions can provide a framework for performing sensitivity analyses to assess the trends in the outcome variable, the effect of treatment or other baseline covariates on the trends or the outcome variables, the role of time-varying covariates, etc.

A general strategy is to create a "wide format" data file with all the variables across all the waves "strung out" with one record per subject in the data file. The SRMI approach is then used to multiply impute the missing values to create M completed-data sets which can then be analyzed using complete data methods.

The next few sections provide several example analyses of longitudinal data with missing covariates and outcome variables with both normal and non-normal outcome variables. Sensitivity analysis is also performed to explore the effect of deviation from the MAR assumption.

8.2   Example 1: Binary Outcome

This example uses data from the American Changing Lives Survey. The study began in 1986 with a national face-to-face survey of 3,617 adults ages 25 and up in the continental U.S., with African Americans and people aged 60 and over over-sampled at twice the rate of the others, and face-to-face re-interviews in 1989 of 83% (n=2,867) of those still alive. Survivors were re-interviewed by telephone, and where necessary, face-to-face in 1994 and 2001/02 and again in 2011/12. (http://www.isr.umich.edu/acl).

The data set used in this example is prepared in a wide format and includes a differently named impairment variable for each wave coded 1=moderate/severe/death and 0 otherwise. For example, I1 represents wave 1=1986, I2=1989, I3=1994, I4=2001/2001, and I5=2011/12. There is missing data on each of the wave 2-5 impairment variables with fully observed data for all other variables such as impairment at wave 1, age, sex, race, and socio-economic status. Overall, about 50% of the cases have item missing data.

The imputation code below demonstrates the use of IMPUTE to impute missing data on the wave 2-5 impairment variables. The IMPUTE call creates 25 multiples each with 20 iterations, use of the TRANSFER statement to include the CASEID variable in the output data set, a SEED value, and use of the DIAGNOSE option to output plots to assist in evaluation of the imputation of the wave 2 impairment indicator.

```
<sas name="Long_Binary">

libname d 'P:\IVEware_and_MI_Applications_Book\DataSets\
ACL Data and Imputation\';
* non imputed wide data set ;
data acl1 ;
 set  d.acl_raw_4sep2016 ;
* examine variables and missing data problem ;
options nofmterr ;
proc means n nmiss mean std min max nolabels ;
run ;
<impute name="bin_impute"> ;
title "Longitudinal Imputation and Analysis" ;
datain acl1 ;
```

```
dataout aclimp all;
default categorical;
continuous v2000;
transfer caseid;
iterations 20;
multiples 25;
seed 67566;
diagnose i2 ;
run;
</impute>
```

The next set of commands demonstrate how to prepare the imputed data for analysis of completed data. For example, a new variable called _imputation_ is set equal to _mult_ for use in analyses to come in SAS, the wide data set is converted to a long or multiple records per individual data set, those age 51+ are deleted to avoid bunching of deaths, and a few additional variables are created for the final analyses. These data management steps are needed so that analyses account for repeated measures for individuals. Specifically, we require the correct data structure and format for use with PROC GENMOD with a REPEATED statement. Lastly, logistic regression output from each imputation multiple is saved and combined using PROC MIANALYZE. We also request multivariate tests of the slopes and intercepts from this procedure through use of the TEST statement and produce an output data set using ODS OUTPUT for post-hoc computation of odds ratios.

```
* Prepare data for analysis ;
data aclimp1   ;
 set aclimp ;
_imputation_ = _mult_ ;
run ;

*Prepare long imputed data set ;
data longacl ;
  set aclimp1 ;
  array  c (1:5) i1 - i5 ;
  do wave = 1 to 5 ;
    impair = c(wave);
    output ;
  end ;
run ;

data longacl1 ;
```

```
  set longacl ;
  if wave=1 then time=0;
  if wave=2 then time=3/10;
  if wave=3 then time=8/10;
  if wave=4 then time=16/10;
  if wave=5 then time=25/10;
  bl_t=bl*time;
  blm_t=blm*time;
  bum_t=bum*time;
  bh_t=bh*time;
  wl_t=wl*time;
  wlm_t=wlm*time;
  wum_t=wum*time;
  wh_t=wh*time;
  if v2000 >50 then delete; * delete those over 50 years ;
run ;


proc sort;
  by _Imputation_;
run ;


proc genmod data=longacl1 desc;
  class caseid wave;
  model impair=v1801 v2000 bl blm bum bh wl wlm wum
  time bl_t blm_t bum_t bh_t wl_t wlm_t wum_t/covb
  dist=bin link=logit;
  repeated subject=caseid /type=un covb printmle;
  by _Imputation_;
  ods output ParameterEstimates=gmparms ParmInfo=gmpinfo
    CovB=gmcovb;
run ;


proc mianalyze parms=gmparms covb=gmcovb parminfo=gmpinfo;
  modeleffects intercept v1801 v2000 bl blm bum bh
    wl wlm wum
      time bl_t blm_t bum_t bh_t wl_t wlm_t wum_t;
  intdiff:test  bl=wl,blm=wlm,bum=wum, bh/mult;
  slopediff:test bl_t=wl_t,blm_t=wlm_t,
  bum_t=wum_t,bh_t/mult;
  ods output  testparameterestimates =outtests ;
run;


proc print data=outtests ;
run ;
```

```
data OR ;
 set outtests ;
 or=exp(estimate) ;cilow=exp(estimate - (1.96*stderr)) ;
  ciup=exp(estimate  + (1.96*stderr)) ;
run ;
title1 "intdiff:test  bl=wl,blm=wlm,bum=wum, bh/mult; " ;
title2 "slopediff:test bl_t=wl_t,blm_t=wlm_t,
bum_t=wum_t,bh_t/mult;" ;
proc print noobs ;
var test or cilow ciup ;
</sas>
```

Table 8.1: Comparison of Intercepts and Slopes Between African-Americans (AA) and Whites (W) in Four Socio-Economic Groups, Source: Imputed ACL Data

SES Status	AA vs W: Baseline OR (95% CI)	AA vs W: OR(per decade) (95% CI)
Low	1.469 (0.942, 2.291)	1.142 (0.819, 1.593)
Lower Middle	2.173 (1.448, 3.261)	1.072 (0.819, 1.403)
Upper Middle	2.189 (1.434, 3.342)	1.112 (0.829, 1.491)
High	3.452 (1.376, 8.658)	1.065 (0.569, 1.994)

Based on results from Column 1, Table 8.1, African-Americans have higher baseline prevalence of impairment, as compared to Whites, for each socio-economic group. Also from Table 8.1,Column 2 or Odds Ratios per decade, shows higher rates of change in prevalence of impairment for AA v. Whites, though none are significant. These results suggest that African-Americans fare worse than Whites when evaluating prevalence of impairment by SES and change over the decades of this study.

8.3   Example 2: Continuous Outcome

The second example uses data from the Opiod detoxification study. The data is described in detail in Appendix A. Our analytic interest is the impact of the treatment drug, buprenorphine-naloxone (Bup-NX), on daily visual analog scores (VAS, range 0 to 100). The comparison is to respondents that received the drug Clonidine. There are 113 individuals with 15 measurements, the first is at baseline and 14 more obtained on a daily basis during the trial. Due to missing data on the outcome, VAS, imputation using a Complete as Randomized approach is demonstrated.

The code below demonstrates data management and data structuring for evaluation of the missing data problem, use of IMPUTE to impute missing data on the 14 outcome variables (using a wide format data set and M=25 with bounds applied for each outcome variable), conversion of the imputed wide data set to a long file appropriate for longitudinal data analysis, and use of SAS PROC MIXED, PROC MIANALYZE, and PROC SGPLOT for analysis of the impact of treatment on the daily VAS measurements. The code is presented in full prior to results/discussion.

```
<sas name="Long_Continuous">
/* Example 2: Opioid Data Set, completed as randomized example*/
libname d3 'P:\IVEware_and_MI_Applications_Book
\DataSets\Longitudinal Data Set';


/* means for wide data set */
proc means data=d3.od_all n nmiss mean min max ;
 class treat ;
run ;
/* impute missing data on outcomes*/
<impute name="od_imp">
 title "Opioid Detox Impute" ;
 datain d3.od_all;
 dataout odimp all;
 default continuous ;
 categorical female white ;
 transfer instudy usubjid ;
 bounds vas0 (>=0, <=100) vas1 (>= 0, <=100)
 vas2 (>=0, <=100) vas3 (>= 0, <=100) vas4 (>=0, <=100)
  vas5 (>= 0, <=100)
vas6 (>=0, <=100) vas7 (>= 0, <=100) vas8 (>=0, <=100)
vas9 (>= 0, <=100)  vas10 (>=0, <=100) vas11 (>= 0, <=100)
vas12 (>=0, <=100) vas13 (>= 0, <=100)  vas14 (>= 0, <=100);
 iterations 5 ;
 multiples 25 ;
 seed 2017 ;
run;
</impute>

/* analyze imputed data sets and set _mult_ to
_imputation_ for use in SAS */
data casrimpw;
 set odimp ;
 _Imputation_=_mult_;
 drop _mult_;
```

```
run ;

/* create a long format data set for subsequent analysis */
data casrimpl;
 set casrimpw;
 array avas(0:14) vas0 - vas14 ;
 do day = 0 to 14 ;
   vas = avas(day);
  output ;
 end ;
 drop vas0 - vas14 ;
run ;

/* each id now has 25*15 records or
113*25*15=42375 records */
proc print data=casrimpl (obs=375) ;
run ;

/* analyze long and imputed data Complete as Randomized */
data casrimpl_analyze_mac ;
 set casrimpl;

/* create a series of variables for final model */
day2=day*day;
tday=treat*day;
tday2=treat*day*day;

/* 14 day dummys plus treat*day */
%macro dum ;
 %do i=1 %to 14 ;
  day&i=0 ;
  if day=&i then day&i=1 ;
  tday&i=treat*day&i ;
 %end ;
%mend dum ;
%dum ;
run ;
/* sort by _imputation_*/
proc sort;
  by _imputation_;
run ;

/* PROC MIXED model with VAS predicted by covariates
 of interest*/
proc mixed data=casrimpl_analyze_mac dfbw;
```

```
  class usubjid;
  model vas=age female white treat day1 day2 day3
  day4 day5 day6 day7 day8 day9 day10 day11 day12 day13 day14
  tday1 tday2 tday3 tday4 tday5 tday6 tday7 tday8 tday9
  tday10 tday11 tday12 tday13 tday14/s covb;
  repeated /subject=usubjid type=un;
  by _Imputation_;
  ods output SolutionF=mixparms covb=mixcovb;
run;


proc mianalyze parms=mixparms covb(effectvar=rowcol)=mixcovb;
 modeleffects intercept age female white treat
 day1 day2 day3 day4 day5 day6 day7 day8 day9 day10 day11
 day12 day13 day14
 tday1 tday2 tday3 tday4 tday5 tday6 tday7 tday8 tday9
 tday10 tday11 tday12 tday13 tday14;
 ods output parameterestimates=outest ;
run;

/* use PROC PRINT and PROC SGPLOT to assess impact of
 treatment by day of study*/
proc print data=outest ;
run ;

proc format ;
 value $pf 'treat'='0' 'tday1'='1' 'tday2'='2'
  'tday3'='3' 'tday4'='4'
  'tday5'='5' 'tday6'='6'  'tday7'='7' 'tday8'='8' 'tday9'='9'
  'tday10'='10'
 'tday11'='11' 'tday12'='12' 'tday13'='13' 'tday14'='14' ;
run;

proc sgplot data=outest ;
title "Estimated Mean Daily Difference on VAS,
 Bup-Nx v. Clonidine Groups" ;
label estimate="Estimated Mean Difference" parm='Day'   ;
where parm in ('treat', 'tday1', 'tday2', 'tday3',
'tday4', 'tday5', 'tday6',
'tday7', 'tday8', 'tday9', 'tday10', 'tday11',
 'tday12', 'tday13','tday14') ;
series x=parm y=estimate;
format parm $pf. ; yaxis min=-15 max=5 ;
run ;
</sas>
```

Table 8.2: Covariate-adjusted MI Mean Difference Between Bup-NX and Clonidine Groups for Each Study Day, Standard Error and p Value

Day	Estimate	SE	p Value
0	-4.216	6.076	0.488
1	1.175	4.989	0.814
2	-9.328	7.825	0.233
3	-9.150	8.182	0.264
4	-6.577	8.044	0.414
5	-6.144	8.553	0.473
6	2.046	7.506	0.785
7	2.464	7.626	0.747
8	-2.286	7.685	0.766
9	-8.014	7.382	0.278
10	-12.842	7.888	0.105
11	-6.974	7.534	0.355
12	-2.764	9.411	0.770
13	-3.904	8.933	0.663
14	-9.570	9.719	0.327

Results from Table 8.2 suggest that those in the treatment (Bup-NX) group report fewer cravings for opioid, (based on the VAS outcome), as compared the control group (Clonidine), though none are statistically significant.

Figure 8.1 presents estimated daily mean differences between the two groups for the 14 day period. Overall, a downward trend is observed with a number of dips and upward trends over the 14 days.

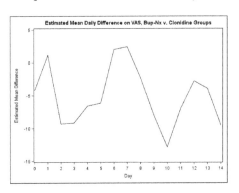

Figure 8.1: Estimated daily mean difference between Bup-NX and clonidine groups

8.4 Example 3: A Case Study

Example 3 is a case study that demonstrates a step-by-step guide to data management, imputation, and subsequent analysis of completed data using a combination of SAS and *IVEware* to correctly analyze multiply imputed complex sample data using multi-level mixed models.

The analytic goal is to estimate household head's wages/salary over time using linear mixed models to account for repeated measures and random effects as well as complex sample design features and the variability introduced by multiple imputation. Use of SAS PROC MI, PROC MIXED, along with the SASMOD and IMPUTE commands of *IVEware* is demonstrated. This example uses individual level data merged with family variables and downloaded from the Panel Study for Income Dynamics (PSID) data center, (http://psidonline.isr.umich.edu). Data set-up statements (for SAS) and raw data were provided by the data center.

The analysis uses the household head's wages/salary from the previous calendar year as the dependent variable along with gender and sample type as covariates. However, prior to imputation/analysis, cases with imputed values on head's wages/salary (done by PSID staff using primarily "hotdeck" methods) were set back to missing for the demonstration of imputation using the Sequential Regression (SMRI) method.

The following code includes four sections: 1. data management before imputation, 2. imputation of missing data and evaluation of imputation using diagnostic tools, 3. data shaping to produce a "long" data set, and 4. analysis of completed data sets using a variety of SAS and *IVEware* commands for descriptive and regression analyses.

Part 1 of the code first calls a working data set organized in a wide format and selects individuals who were household heads and in the PSID family in each year 1997, 1999, 2001, 2003, 2005, 2007, 2009, 2011, 2013 and part of the original 1968 PSID sample defined as from either the Survey Research Center (SRC) or Census sample. The SRC and Census samples are considered "core sample" and have been followed from 1968 to the present, see the PSID documentation for more details on the PSID sampling methods. The 2013 PSID individual longitudinal weight is used for all analyses as it includes attrition adjustments over the years of interest, 1997-2013 (see PSID documentation for details on weight construction) and because all individuals were present in each year. Other weighting options might be considered but for simplicity we use the last weight of the years considered.

The head's wages/salary variables are used in their original scale (dollar amounts in year of collection) and converted to 2013 dollars post-imputation for final analyses. Other variables included are head's grade of completed education, age, gender, plus individual longitudinal weight and complex sample design variables required for correct design-based estimation. Note that each of these constructs have nine individually named variables for each record.

Part 2 performs imputation of missing data for the nine head's wages/salary and nine education variables. Since there is a relatively small amount of missing data, five imputation multiples are generated. The wide data structure is recommended to ensure that within individual relationships are maintained during the imputation. In addition, use of a BY statement in IMPUTE requests imputation be done separately within each sample (SRC and Census). Other options are inclusion of a combined categorical stratum and cluster variable to account for the complex sample design, the 2013 individual longitudinal weight, and additional model covariates to enrich the imputation. Diagnostic plots of observed v. imputed values are produced by both *IVEware* (diagnose option) and PROC SGPLOT.

Part 3 of the command syntax restructures the five completed data sets into a long data set suitable for longitudinal data analysis. Arrays are used with an output statement to generate a set of variables that each represent each construct at 9 time points and wages/salaries are converted to 2013 dollars (with Consumer Price Index conversion factors) during this step.

Part 4 code calls the data set produced in Part 3 to perform analyses such as an unconditional means model stratified by sample category and year, a plot of a few individual records over time to show individual wage/salary variation, and a linear growth model predicting head's wages/salary using random intercepts/slopes, an interaction between time and sample while controlling for gender. Multi-level modeling is performed with *IVEware* SASMOD and PROC MIXED since the combination of commands offers Jackknife Repeated Replication to account for the complex sample features as well as MI combining implemented automatically.

8.4.1   Code

```
sas name="PSID Case Study">
/* Part 1 */
/* Chapter 12 Linear Growth Model, PSID data 1997-2013*/
/* Analysis focuses on individuals that were
        1. heads in each year 1997-2013, (relationship to
        head is 10 for each year)
        2. and from either SRC or Census samples in 1968,
        3. and sequence number 1-20 for each year

*/

libname d 'P:\IVEware_and_MI_Applications_Book
\Chapter12Simulations\Examples\PSID data' ;

/* data j221871 refers to PSID job number produced by
```

```
PSID data center extract system*/
data anal ;
set d.j221871;
if
(1<=er30001<=2930 or 5001<=er30001<=6872)
  and er30002 <=169
  and (er33403=10 and 1<=er33402<=20)
  and (er33503=10 and 1<=er33502<=20)
  and (er33603=10 and 1<=er33602<=20)
  and (er33703=10 and 1<=er33702<=20)
  and (er33803=10 and 1<=er33802<=20)
  and (er33903=10 and 1<=er33902<=20)
  and (er34003=10 and 1<=er34002<=20)
  and (er34103=10 and 1<=er34102<=20)
  and (er34203=10 and 1<=er34202<=20)
;

length samplecat $6 ;
if 1<=er30001<=2930 then samplecat='SRC' ;
else samplecat='Census' ;

* set 99 on completed education to missing and impute ;
array ed [*] er33415 er33516 er33616 er33716 er33817
 er33917 er34020 er34119 er34230 ;
array edr [*] ed1-ed9 ;

* set cases imputed by PSID back to missing
 (values of 1,2,3,4,5) on accuracy variables
  for head wages ;
array ac [*] er12197 er16494 er20426 er24118 er27914
 er40904 er46812 er52220 er58021 ;

* prepare new string of head wages with imputations removed ;
array wa [*] hdwg1-hdwg9 ;
array wa1 [*] er12196 er16493 er20425 er24117 er27913
 er40903 er46811 er52219 er58020 ;
* create age in each of 9 waves from age in 1968 ;
array age [*] er33404   er33504 er33604 er33704 er33804
 er33904 er34004 er34104 er34204 ;
array ag [*] age1-age9 ;

do i = 1 to 9 ;
  if ed[i]=99 then edr[i]=. ; else edr[i] = ed[i] ;
  if ac[i] in (1,2,3,4,5) then wa [i] = . ;
   else wa[i]=wa1[i] ;
```

```
  ag[i]=age[i] ;
end ;

* labels for new variables for imputation example ;
label
ed1='Yrs Completed Ed 1997'
ed2='Yrs Completed Ed 1999'
ed3='Yrs Completed Ed 2001'
ed4='Yrs Completed Ed 2003'
ed5='Yrs Completed Ed 2005'
ed6='Yrs Completed Ed 2007'
ed7='Yrs Completed Ed 2009'
ed8='Yrs Completed Ed 2011'
ed9='Yrs Completed Ed 2013' ;

label
hdwg1='Head Wages 1996'
hdwg2='Head Wages 1998'
hdwg3='Head Wages 2000'
hdwg4='Head Wages 2002'
hdwg5='Head Wages 2004'
hdwg6='Head Wages 2006'
hdwg7='Head Wages 2008'
hdwg8='Head Wages 2010'
hdwg9='Head Wages 2012' ;

label
age1='Age 1997'
age2='Age 1999'
age3='Age 2001'
age4='Age 2003'
age5='Age 2005'
age6='Age 2007'
age7='Age 2009'
age8='Age 2011'
age9='Age 2013' ;
run ;

/* Part 2 */
data psid1 ;
 set anal ;
 *create imputed value flags for use in diagnostics;
 if ed1=. then imped1=1 ; else imped1=0 ;
 if hdwg1=. then imphdwg1=1 ; else imphdwg1=0 ;
 if hdwg2=. then imphdwg2=1 ; else imphdwg2=0 ;
```

```
* create a combined stratum + psu variable ;
strat_psu = (er31996 *100) + er31997 ;


* create a unique id er30001 + er30002 ;
id=er30001*1000 + er30002 ;

keep
er30001 er30002 id  strat_psu er31996 er31997 samplecat
age1-age9 er32000 ed1-ed9 hdwg1-hdwg9
er33430 er33546 er33637 er33740 er33848 er33950 er34045
 er34154 er34268
imped1 imphdwg1 imphdwg2 ;
run ;


/* impute missing data on head wages 1996-2012, education
1997-2013 using gender, age, design variables and weights*/
<impute name="impute_psid_by">
title "Impute PSID Missing Data" ;
datain  psid1 ;
dataout impute_psid_ive all ;
default continuous ;
transfer id er30001 er30002 imped1 imphdwg1 imphdwg2
samplecat er31996 er31997 ;
categorical er32000 ed1 ed2 ed3 ed4 ed5 ed6 ed7 ed8 ed9;
by samplecat ; *impute using by statement for sample
categories ;
*set bounds for wages ;
bounds
 hdwg1 (>=0, <=1000000) hdwg2 (>=0, <=1000000)
 hdwg3 (>=0, <=1000000) hdwg4 (>=0, <=1000000)
 hdwg5 (>=0, <=1000000)
 hdwg6 (>=0, <=1000000) hdwg7 (>=0, <=1000000)
  hdwg8 (>=0, <=1000000)
 hdwg9 (>=0, <=1000000) ;
 iterations 5;
 multiples 5;
 seed 2017;
 diagnose hdwg2 ;
run ;
</impute>


*plot with 4 categories as different symbols ;
data impute_psid_ive1 ;
```

```
  set impute_psid_ive ;

*compare headwage1 v. headwage2 ;
 if      imphdwg1=1 and imphdwg2=1 then cat1='II' ;
 else if imphdwg1=1 and imphdwg2=0 then cat1='IO' ;
 else if imphdwg1=0 and imphdwg2=1 then cat1='OI' ;
 else if imphdwg1=0 and imphdwg2=0 then cat1='OO' ;
run ;

proc sgplot data=impute_psid_ive1 ;
 where _mult_=1 ;
 scatter y=hdwg1 x=hdwg2 /group = cat1 ;
run ;

/* Part 3 */

/* Create a long data set with multiple records per person
 within each multiple, using completed data sets*/
data long_imputed_ive ;
 set impute_psid_ive ;
 if samplecat='SRC' then src=1 ; else src=0 ;
 if er32000=2 then female=1 ; else female=0 ;
 age97=age1 ;

 * use arrays to output multiple records per individual ;
 array w [*] hdwg1- hdwg9 ; * head wages for each year ;
 array ed [*] ed1-ed9 ; * completed education for each year ;
 array y [9]   _temporary_ (1997 1999 2001 2003 2005 2007
  2009 2011 2013) ; * create year of study ;
 array wi [9] _temporary_ (1.45 1.40 1.32 1.27 1.19 1.12
  1.09 1.04 1.00) ; * convert each yr to 2013 dollars ;
 array weight [*] er33430 er33546 er33637 er33740
 er33848 er33950 er34045 er34154 er34268 ; * weights ;
 array ag [*] age1-age9 ;  * age in each yr ;

do i = 1 to 9 ;
 hdwg= w[i];
 headwage=hdwg * wi[i] ;
 wgt=weight[i];
 sex=er32000;
 stratum=er31996;
 cluster=er31997;
 age = ag[i];
 year = y[i];
 time = i-1 ;
```

```
 completeded=ed[i] ;
 output ;
end ;

keep hdwg headwage wgt sex stratum cluster age
age97 year time completeded _mult_ er30001 er30002
 er32000 samplecat er34268 id;
run ;

proc sort data=long_imputed_ive ;
 by _mult_ ;
run ;

/* Part 4 */

/* Unconditional Means Model by Sample Category and Year */
<sasmod name="Unconditional Means Model with Design
Adjustments and Wgt IVEWARE">
title SASMOD Unconditional Means Model Head Wages by
Sample for 1997 to 2013 ;
datain long_imputed_ive ;
cluster stratum ;
stratum cluster ;
weight  er34268 ;
/* SAS statements begin here */
proc mixed ;
 class year samplecat ;
 model headwage = year*samplecat / noint solution ddfm=bw;
 repeated / subject=id ;
run ;
</sasmod>

* Use means from above model for plotting ;

data means ;
input Year Sample $ Estimate  Se;
datalines ;
1997 Census      25821.8349808    1655.1656610
1997 SRC         50121.3210007    2096.9368941
1999 Census      28599.8266492    1793.1101554
1999 SRC         52692.2588232    2158.9495567
2001 Census      33536.8828587    4609.5273727
2001 SRC         54338.0924074    2577.3028706
2003 Census      28266.6584767    1907.7950353
2003 SRC         52632.4814954    3776.2446517
```

```
2005 Census        29202.0905887        2556.2529193
2005 SRC           54195.5643616        3034.8029945
2007 Census        26977.7090998        2297.8630857
2007 SRC           52992.3306134        2400.5511389
2009 Census        27897.7098894        2382.9680033
2009 SRC           52285.2503362        3223.9304476
2011 Census        22184.2218272        2337.8229310
2011 SRC           43681.4480980        2330.1307490
2013 Census        21632.9758473        2220.3276362
2013 SRC           48358.8517553        5374.0000000
;
run ;


/* Plot Means by Sample per Year*/
title "Plot of Mean Head Wage/Salary by Year" ;
proc sgplot data=means ;
 series x=year y=estimate / group=sample markers ;
 xaxis label='Year' ; yaxis label='Mean Head Wage/Salary
   1997 to 2013 (in 2013 Dollars)' ;
run ;


/* Prelude to Linear Growth Model: Individual Lines to Show
Variation Between Individuals */
proc sgplot data = long_imputed_ive ;
   series x=year y=headwage / group=id markers ;
   where _mult_=1 and id in (7035, 2524001, 6872003, 6845006)  ;
   xaxis label='Year 1997 - 2013 Odd Years' ; yaxis label='Head
    Wage/Salary' ;

run;



/* Linear Growth Model */
<sasmod name="Linear Growth Model with Time
SampleCat and Gender">
title SASMOD Growth Model from PROC MIXED;
datain long_imputed_ive ;
cluster stratum ;
stratum cluster ;
weight   er34268 ;
proc mixed ;
 class id samplecat er32000 ;
 model headwage = time samplecat time*samplecat er32000
 /solution ddfm=bw ;
 random intercept time / type=un subject=id ;
```

```
run ;
</sasmod>

* Predicted Head Wages using Mixed Model Results ;
data predicted;
   set long_imputed_ive ;
 predicted_hdwage=
   32411.39 +
   time * -670.29 +
   (samplecat='Census') * -15734.37 +
   (samplecat='SRC') * 0  +
   time*(samplecat='Census')* 194.05 +
   time*(samplecat='SRC')* 0 +
   (er32000=1)*24897.39 +
   (er32000=2)*0 ;
   if er32000=1 then gender='M' ; else gender='F' ;
   length samplecat_gender $55 ;
   samplecat_gender=trim(samplecat)||'_'||(gender) ;
run ;

proc means data=predicted ;
  var predicted_hdwage ;
  class samplecat year ;
  weight er34268 ;
  ods output summary = outstat ;
run ;

/* By Samplecat */
proc sgplot data=outstat ;
 label samplecat='Sample Categories' ;
 series x=year y=predicted_hdwage_mean /
 group=samplecat markers ;
 xaxis label='Year of Interview' ;
 yaxis label='Predicted Mean
  Head Wage/Salary 1997 to 2013 (in 2013 Dollars)' ;
run ;

/* By Samplecat and Gender */
proc means data=predicted ;
  var predicted_hdwage ;
  class samplecat_gender year ;
  weight er34268 ;
  where age >=30 and age <=65 ;
  ods output summary = outstat1 ;
run ;
```

```
proc sgplot data=outstat1 ;
 label samplecat_gender='Sample Categories and Gender' ;
 series x=year y=predicted_hdwage_mean /
 group=samplecat_gender markers ;
 xaxis label='Year of Interview'  ; yaxis label='Predicted
  Mean Head Wage/Salary 1997 to 2013 (in 2013 Dollars)' ;
run ;
</sas>
```

8.4.2   Analysis Results

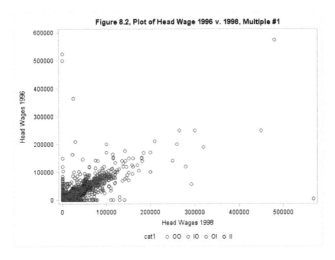

Figure 8.2: Head Wages/Salary (1998) by Observed v. Imputed

Figures 8.2 and 8.3 show plots of imputed v. observed values for head's wages/salary for year 1998 and a similar plot for wages/salary 1996 by 1998. Neither of these diagnostic plots reveal issues/concerns about the quality of the imputations given that the observed and imputed values distributions appear to be similar. (See Part 2 syntax for details).

Table 8.3 and Figure 8.4 show tabular and graphical representations of weighted means and standard errors, for head's wages/salary by year/sample. These statistics are derived from PROC MIXED and SASMOD using an unconditional means model that accounts for repeated measures of individuals (within-subject variation), complex sample design features, and the increased variability due to multiple imputation. The results suggest that SRC sample respondents have much higher mean wages/salary as compared to the Census sample while the trend over time shows rising mean wages/salaries until 2001 followed by a downward trend from 2003 to 2013. All values are in 2013 dollars

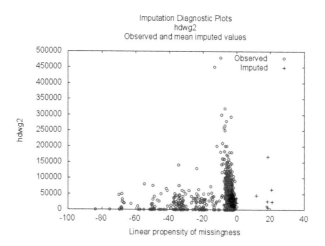

Figure 8.3: Head Wages/Salary(1996) v. Head Wages/Salary(1998), Observed v.Imputed

Table 8.3: Estimated Means and Standard Errors for Head Wages/Salary 1997-2013 by Year and Sample

Year	Sample	Mean (SE)	Sample	Mean(SE)
1997	Census	25821 (1655)	SRC	50121 (2097)
1999	Census	28600 (1793)	SRC	52692 (2159)
2001	Census	33537 (4609)	SRC	54338 (2577)
2003	Census	28267 (1908)	SRC	52632 (3776)
2005	Census	29202 (2556)	SRC	54196 (3035)
2007	Census	26978 (2298)	SRC	52992 (2401)
2009	Census	27898 (2383)	SRC	52285 (3224)
2011	Census	22184 (2338)	SRC	43681 (2330)
2013	Census	21633 (2220)	SRC	48359 (5374)

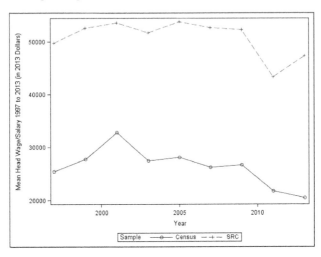

Figure 8.4: Head's Wages/Salary by Year and Sample

and as a reminder, represent dollar amounts from the previous year. (See Part 4 of code for details).

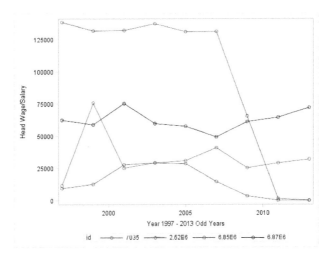

Figure 8.5: Individual Trends of Head Wages/Salary by Year

Figure 8.5 illustrates trends over time for 4 unique individuals and highlights how each respondent has a different intercept and slope for the outcome of interest over time. This serves as background for the next analysis which demonstrates use of random intercepts and slopes in a linear growth model with covariates. (see code, Part 4).

Table 8.4 presents parameter estimates, standard errors, Wald tests, and *p*

Table 8.4: Parameter Estimates of Head Wages/Salary from Linear Regression with Random Intercepts and Slopes

Variable	Estimate	SE	Wald	Prob > Chi
Intercept	32411.392	2150.473	227.157	0.000
Time	-670.298	325.168	4.249	0.039
Census	-15734.372	2137.028	54.209	0.000
Time*samplecat (Census)	194.052	341.493	0.323	0.570
ER32000 1(Male)	24897.390	1866.345	177.961	0.000

values from a linear growth model predicting head wages/salary using PROC MIXED and SASMOD. The mixed model specifies random intercepts and slopes while using sample, sample*time, and gender as predictors. Note that time is treated as a continuous predictor in this model while sample (Census=1, SRC=0) and gender (Male=1, Female=0) are considered dummy variables.

The results suggest that sample and gender are both significant predictors of head's wage/salary over time but the non-significant interaction indicates a common slope is appropriate for this model. Again, use of PROC MIXED/SASMOD permits design-adjusted variance estimates for a linear growth model that also employs correct MI combining rules. The model estimates are then used to calculate predicted values and additional analyses/plots are produced using the predicted values.

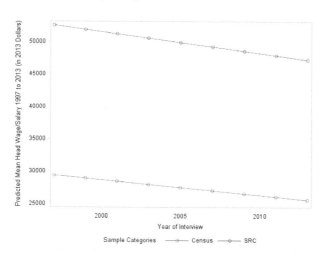

Figure 8.6: Predicted Wages/Salary by Sample, among all ages

Figures 8.6 and 8.7 use predicted values based on the growth model estimates and allow examination of trends in wages/salaries over time. Figure 8.6 presents the predicted wages/salaries for all ages while Figure 8.7 is restricted

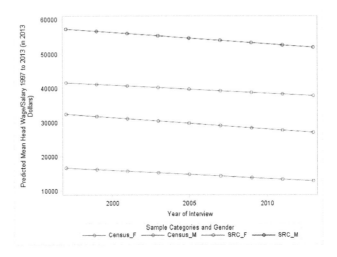

Figure 8.7: Predicted Wages/Salary by Sample and Gender, among those age 30-65

to those aged 30-65 to cover a broad age range of working adults. The plots suggest a small decrease in head's wages/salary between 1996-2012 with about a \$25,000 difference between the SRC and Census samples. This same trend is also noted when looking at predicted head's wages/salary stratified by sample and gender. For example, based on Figure 8.7, males earn more than women from both samples. These results highlight a negative trend over time, income differences between the two PSID samples as well as the persistent gender gap in wages/salary in the US during 1996-2012.

8.5 Discussion

The multiple imputation approach used in this chapter uses an unstructured mean and the covariance matrix to impute the missing values. However, there may be some structural constraints that needs to be incorporated in the imputation process. Consider a study of young adolescents, where Z_i is a vector of baseline covariates and $Y_{it} = (Y_{1it}, Y_{2it}, \ldots, Y_{pit})$ is a vector of variables measured at time $t = t_1, t_2, \ldots, t_T$ on the adolescent subject $i = 1, 2, \ldots, n$, and an arbitrary pattern of missing data. Suppose that one of the variable, Y_1, is height. Clearly, for any $t_k < t_{k+1} < t_{k+2}$, it is reasonable to expect the constraint, $Y_{1it_k} \leq Y_{1it_{k+1}} \leq Y_{1it_{k+2}}$ to be met for every subject $i = 1, 2, \ldots, n$. Such constraints can be incorporated in *IVEware* using the bounds statement.

Monotonicity constraints among the observations for each variable should be carefully considered and incorporated as additional structures.

The second type of structure involves a temporal pattern. For example, the variable Y_j may satisfy a structure,

$$Y_{ijt} = f_j(\alpha_i, t) + \epsilon_{ijt}$$

where f_j is a known function of time t involving parameters specific to the subject i, α_i. The residuals, ϵ_{ijt}, and the subject specific parameters, α_i, are treated as random with certain distributions. Multiple imputation procedures that incorporate the temporal structure may be more efficient than the unstructured mean and covariance matrix considered in this chapter. Incorporation of such structures involve tailored programming and needs to be considered on the case-by-case basis. The view taken (in this book) is that the unstructured mean and covariance are perhaps the most general purpose approach when multiple variables of different types are collected at each time point, in addition to the baseline covariates, with an arbitrary pattern of missing data and when the goal is to borrow as much strength as possible from the other observed variables on each subject to predict the missing values.

8.6 Additional Reading

A common method in longitudinal clinical trials is the LOCF (last observation carried forward) approach. A review of this and other early approaches can be found in Heyting, Tolboom and Essers (1992). See Gadbury, Coffey and Allison (2003) and Engels and Diehr (2009) for an overview from an applied perspective. Lavori, Dawson and Shera (1995) proposed multiple imputation methods. The maximum likelihood approach using the EM-algorithm is discussed in Laird and Ware (1982) and Dempster, Rubin and Tsutakawa (1981). Little (1995) provides a lucid discussion of modeling issues with drop outs in the longitudinal studies. Little and Rubin (2002) also provide extensive discussion of the maximum likelihood estimation for longitudinal data. Siddiqui and Ali (1998) and Mallinckrodt et al (2001, 2003) compare various approaches for analyzing incomplete data. Diggle, Heagerty, Liang and Zeger (2002), Verbeke and Molenberghs (2000) and Molenberghs and Verbeke (2005) provide comprehensive coverage of issues in the analysis of longitudinal data including an extensive discussion of missing data. Another good read is the report by National Research Council (2010).

The inverse probability weighting and general semiparametric approaches for analyzing repeated measures data with missing values are discussed in Robins, Rotnitzky and Zhao (1995). See also Davidian, Tsiatis and Leon (2005). Preisser, Lohman and Rathouz (2002) assess the performance of weighted estimating equation, especially with respect to mispecification of the missing data mechanism. The primary reference for Bup-Nx and clonidine comparison trial is Ling et al (2005) and for the American Changing Lives (ACL) is the ICPSR archive, House (2014).

8.7 Exercises

1. Download the data set called **Ex_longitudinal** from the book web site. This data comes from a study of epileptic seizures and is used to practice methods for multiple imputation and analysis of longitudinal/repeated measures data, see Liang and Zeger, (1986) for details.

   The data set consists of counts of the number of seizures experienced in each of four 2 week periods during the trial; the baseline number of seizures for the 8 week period prior to the start of the clinical trial; subject ID, age in years; an indicator of whether the subject received an anti-epilepsy drug treatment or placebo (control); and a time variable representing the four 2 week periods.

   For this exercise, we modified the original data such that there is missing data for one of more of the four seizure count variables. There are 59 unique individuals in the study, each with 4 data records, one for each of the 4 two week observation periods. The structure of the modified data set is a multiple records per respondent rectangular data array. The intended analysis focuses on a regression of number of seizures experienced during the trial on the covariates: baseline seizures, age, treatment status, and time (representing the four measurements during the trial).

   (a) Using your software of choice, examine the extent and pattern of missing data and take note of the multiple records per individual data structure. What variables have missing data and fully observed data? How much missing data is present? Why might you restructure the data for imputation?

   (b) Reshape the long data set into a wide data set with one record per individual and include new variables containing count of seizures at each time point 1-4.

   (c) Impute missing data using the IMPUTE command. Create M=10 data sets with 10 iterations per multiple, use a SEED value of 45678, and make sure to correctly declare each variable type.

   (d) Restructure the concatenated imputed data sets to a long format (for data analysis) with an output statement or similar technique depending on your software that produces 4 records per respondent, per imputation, that is, 10 imputations*59 unique people*4 seizure measurements =2,360 records.

(e) Run a Poisson regression model using SAS PROC GENMOD (or similar procedure if using other statistical software) with a repeated statement to account for the lack of independence among respondents. Your repeated measures Poisson model should regress the count of seizures on age, treatment status, time period of seizure measurements, and baseline number of seizures.

(f) Based on your results from (e), outline the data management steps required, the imputation process used, and interpret the regression coefficients/rates. How does the data structure used promote maintaining relationships of seizures per time period/individual?

2. Repeat the second example demonstrated in this chapter, using the Opiod Detoxification data set, but use the "Completed as Control" method previously discussed in this chapter.

   (a) Follow all data management steps as presented above, use M=20 imputations, and be sure to create time-varying treatment variables with missing data assigned to the control group.

   (b) Prepare a table of results similar to Table 8.2 and discuss how the results compare to those where Completed as Randomized is the chosen method.

   (c) Prepare a plot similar to Figure 8.2 in your software of choice and again, describe how these results compare to Completed as Randomized.

3. Generate 500 data sets each with $n = 200$ subjects and 8 observations under the following model assumptions:

   (1) $Z_1 \sim N(1, 2)$

   (2) $Z_2 | Z_1 \sim$ Bernoulli$(1, \pi(Z_1))$ where $\pi(Z_1) = [1 + \exp(-1 - 0.5Z_1)]^{-1}$

   (3) $Y_{it} = \beta_{0i} + \beta_{1i}t + \beta_{2i}\log(t) + \epsilon_{it}$ with $i = 1, 2, \ldots, n = 200$ and $t = 0, 1, 2, 3, 4, 5$.

   (4) $\epsilon_{it} \sim N(0, 1)$ and

   $$\begin{pmatrix} \beta_{0i} \\ \beta_{1i} \\ \beta_{2i} \end{pmatrix} \sim \text{BVN}\left[\begin{pmatrix} 1 + Z_1 + Z_2 \\ 0.5 + 0.5Z_1 + 0.5Z_2 \\ 0.5 - 0.25Z_1 - 0.25Z_2 \end{pmatrix}, \begin{bmatrix} 1 & -0.5 & -0.1 \\ -0.5 & 1 & 0.25 \\ -0.1 & 0.25 & 1 \end{bmatrix} \right]$$

   (5) Missing data mechanisms: No missing values in Y_0, Y_3. Missing values in Z_1, Z_2, Y_1, Y_2, Y_4 and Y_5 depends on Y_o and Y_3. Choose

logistic models for missing data mechanisms that results in no
more than 20% missing in each variable and no more than 20%
complete cases.

The parameters of interest are the contrast in the adjusted means,

$$\theta = E(Y_5 - Y_0 | Z_1, Z_2 = 1) - E(Y_5 - Y_0 | Z_1, Z_2 = 0)$$

The analyst procedure for estimating θ is through running a regression of $Y_5 - Y_0$ on (Z_1, Z_2) and using the regresion coefficient for Z_2 in this model as an estimate of θ.

(a) What is the true value of θ under the stated complete data
model assumptions?

(b) Apply the analyst procedure on each of the 500 complete data
sets (before deletion) and assess the sampling properties of the
estimates and compare it to the true value. Assess the mean
square and confidence coverage properties.

(c) Fit the data generating model on each of the 500 complete data
sets (treating all the parameters as unknown), and hence, derive
the estimates of θ. Compare them to the estimates obtained in
(b).

(d) Repeat (b) and (c) on complete-cases (that is, delete subjects
with missing values in Z_1 and Z_2).

(e) Carefully develop the imputation model using the first generated data set with missing values. Using this imputation model,
multiply impute $(M = 25)$ the missing values in each of the remaining 499 data sets and then apply (b) and (c). Combine the
25 imputed data sets to obtain point and interval estimates of θ
from each of the 499 data sets. Assess the sampling properties
of these point and interval estimates.

(f) For the multiply imputed data sets in (e), apply the methods
described in (b) and (c) using the imputed Z's and the unim-
puted Y's.

(g) Based on (a) to (f), write a report summarizing your findings
and recommendations.

(h) Change the values of the parameters in the data generating
model or missing data mechanisms to explore the robustness of
your findings in (g).

9

Complex Survey Data Analysis using BBDESIGN

9.1 Introduction

Most surveys involve complex design features such as stratification, clustering and weighting (for selection, non-response or post-stratification). To the extent that they have predictive power for variables with missing values, they should be included in the imputation process. The second reason to include them is to make sure that the complex survey data analysis of the completed data sets is compatible with both, the observed and imputed values. That is, the imputed values should reflect the same "design properties as the observed values" to make the completed data sets as plausible data sets from the population using the particular complex survey design.

One approach is to use the variables that are involved in the creation of strata, clusters and weights as predictors along with the covariates in the imputation model. The survey weight in the original and/or transformed scale may be included as well. However, in *IVEware*, a nonparametric procedure is used to generate synthetic populations based on the observed data and the design variables. Multiple imputations are then performed on the synthetic populations using SRMI. For this approach, the combining rules are different. This section illustrates an application of this approach using *IVEware*.

The synthetic population is drawn using the approach described in Zhou, Elliott and Raghunathan (2016a, 2016b, 2016c). The following is a brief description of the procedure. Suppose H is the number of strata with c_h clusters sampled from stratum, $h = 1, 2, \ldots, H$. Let n_{jh} be the number of respondents in cluster $j = 1, 2, \ldots, c_h$ from stratum h. Let w_{ijh} be the weight associated with the respondent $i = 1, 2, \ldots, n_{jh}$ and Y_{ijh} be the corresponding vector of survey variables (including the missing values) .

1. For each $h = 1, 2, \ldots, H$, draw $c_h^* = c_h - 1$ clusters using a simple random sample with replacement from the c_h clusters and index them as $j^* = 1, 2, \ldots, c_h^*$. All the elements in the cluster/PSU are taken into the replicate for the selected cluster/PSU. Thus, the first bootstrap sample is obtained.

2. Adjust the weight, $w_{ijh}^* = c_h w_{ijh}/(c_h - 1)$ if the cluster j in stra-

tum h is included in the replicate and set to 0, if the cluster, is not included in the replicate. At this point, the cluster and stratum indices can be dropped as they have been now replicated in the bootstrap sample. Thus, let w_i^* denote the weights for all the elements in the replicate and normalize to size n^*, the number of elements in the replicate.

3. Repeat the above two steps S times to generate additional replicates.

4. For each bootstrap sample, $s = 1, 2, \ldots, S$ apply the Finite Population Bayesian Bootstrap (FPBB) to generate $N - n^*$ non-sampled elements in the population. Operationally, a weighted Polya urn model is used to draw $N - n^*$ values from the urn containing n^* items (indexed as $A = \{1, 2, 3, \ldots, n^*\}$). The following are the steps:

   (a) Draw a value from $A = \{1, 2, \ldots, n^*\}$ with selection probabilities proportional to $w_i^*, i = 1, 2, \ldots, n^*$. This is the first of $N - n^*$ non-sampled elements in the population.

   (b) Replace the sampled item back into the urn and increase its weight by 1.

   (c) Repeat Step (a) with the revised weights to draw the second item.

   (d) Repeat Step (b).

   (e) Continue until all $N - n^*$ elements have been sampled. The value of N may be chosen to be large relative to n^* (for example, $N = n^*/0.01$) as an approximation.

5. Repeat the FPBB a total of B times, thus, generating $S \times B$ synthetic populations.

6. Missing values in each of the $S \times B$ synthetic populations is multiply imputed M times using SRMI to generate $S \times B \times M$ completed synthetic populations.

Let $\widehat{\theta}_{sbl}$, $s = 1, 2, \ldots, S$; $b = 1, 2, \ldots, B$; and $l = 1, 2, \ldots, M$ be the estimate of the parameter, θ, computed from the completed synthetic population (s, b, l). Let $\widehat{\theta}_{s++} = \sum_b \sum_l \widehat{\theta}_{sbl}/(BM)$. The multiply Imputed estimate is $\widehat{\theta}_{MI} = \sum_s \widehat{\theta}_{s++}/S$ and its variance estimator is

$$V_{MI} = (1 + 1/S) \sum_s (\widehat{\theta}_{s++} - \widehat{\theta}_{MI})^2/(S - 1).$$

The confidence interval is constructed using a t distribution with $\nu = \min(\sum_h c_h - H, S - 1)$ as the degrees of freedom. Note that this approach is computationally intensive and may require several hours to run depending upon the number of synthetic populations S, the number of finite population Bayesian bootstrap samples, B and the number of imputations, M.

9.2 Example

This section uses a case study approach to demonstrate use of the BBDE-SIGN command to implement the weighted finite population Bayesian Boot-strap technique, followed by imputation of missing data, conduct linear and logistic regression analyses of completed data, and, finally, the implementa-tion of the combining rules described in the previous section. The case study uses 2011-2012 NHANES data and focuses on linear regression analysis of to-tal cholesterol predicted by BMI, gender, and family income to poverty ratio along with logistic regression predicting obesity status by age, gender and family income/poverty. The example uses SAS with the SRCShell editor and uses a combination of the SAS data step for the combining steps. These steps can be generalized to other software tools as needed.

9.2.1 Code

The first section of code below calls SAS and gathers selected variables from the NHANES data set and subsets the data to those respondents age 18+ and participated in the medical examination and interview portions of the survey.

The *IVEware* command BBDESIGN is then used to generate 25 "impli-cates" or data sets ($S = 5$ and $B = 5$), using the weighted finite popula-tion Bayesian bootstrap (FPBB) method. The bootstrap uses the Medical Examination Component weight (WTMEC2YR) and complex sample design variables representing the sample stratification and primary sample units (SD-MVSTRA and SDMVPSU) to "uncomplex" the data set. By default, 25 im-plicate data sets are created and can be identified by the internal variable, "_impl_".

The data set total sums to 10 (default multiplier of BBDESIGN)*original n (5,615)* 25 implicates, that is, 10*5,615*25=1,403,750. The output data set is saved as a temporary file called "bbdesignout" via the DATAOUT statement.

```
<sas name="BBDesign Example">

/* BBDesign Example, Uses NHANES 2011-2012 DATA with
 BBdesign and Impute */
libname d 'P:\IVEware_and_MI_Applications_Book
\Chapter12Simulations\Examples\Revised BBDESIGN 12feb2018';

* gather NHANES data where age >=18 and MEC weight > 0
(participated in MEC examination) ;
data nhanes1112_sub_20jan2017  ;
 set d.nhanes1112_sub_4nov2015 ;
 if age >=18 and wtmec2yr > 0 ;
```

```
 drop marcat bpxsy1 - bpxsy4 bp_cat pre_hibp  bpxdi1 - bpxdi4
 dmdmartl irregular ;
run ;


proc means nolabels n nmiss mean min max ;
weight wtmec2yr ;
run ;


/* Use BBDesign command to prepare data set using complex
sample design variables and MEC weight:
 25 implicate data sets are generated:
 5 Bootstrap sample of clusters
 5 FPBB using Weighted Polya posterior within each
  bootstrap sample
*/


<bbdesign name="BBdesign">
 datain nhanes1112_sub_20jan2017 ;
 dataout d.bbdesignout ;
 stratum sdmvstra ;
 cluster sdmvpsu ;
 weight wtmec2yr ;
 csamples 5 ;
 wsamples 5 ;
 seed 2001;
run;
</bbdesign>


/* Confirm that there are 10 (sample inflation factor)*5,615
 (original n) *25 (implicates)= 1,403,750 */
proc freq data=d.bbdesignout ;
 tables _impl_ ;
run ;
```

Next, missing data is addressed via use of the IMPUTE command with M=5. Since the data set has already been prepared to represent the population of interest, we impute missing data values for total cholesterol, family income/poverty ratio, BMI, and education within each implicate. Once this is complete, data analysis can be done using simple random sample assumptions, that is, without use of complex sample design variables or probability weights.

```
/* impute missing data within each of 25 implicates
 using M=5 and 5 iterations */
<impute name="Impute_BBDesign"> ;
 datain d.bbdesignout ;
 dataout d.imputed_samples all ;
 default continuous ;
 transfer ridstatr seqn ag1829 ag3044 ag4559 ag60 mex
 othhis white black other _impl_ _obs_ ;
 categorical riagendr ridreth1 edcat ;
 bounds indfmpir (>= 0, <=5) bmxbmi (>=13, <=80)
 lbxtc (>=59, <=523) ;
 by _impl_;
 seed 2016 ;
 multiples 5;
 iterations 5;
 run ;
</impute> ;
```

Two analyses are now demonstrated; one using a linear regression model and another using logistic regression with combining for both models. This is needed because the default combining rules implemented in REGRESS and PROC MIANALYZE are different from the FPBB rules, see Raghunathan (2016) or Zhou, Elliot and Raghunathan (2016b) for details. The linear regression example uses total cholesterol predicted by gender, BMI, and the ratio of family income to poverty thresholds. For the logistic regression example, a binary outcome of obesity status (coded 1 if BMI $>=30$, and 0 otherwise) is predicted by age, gender and the income/poverty ratio.

```
/* Prepare the imputed synthetic populations for analysis */
data synthpops;
 set d.imputed_samples;
/*
Create 3 indices S, B, L using the fact that wsamples=5 in
the BBDESIGN code above and given the relationships:
****************************************
indexL=_mult_;
indexS=floor((_impl_-1)/wsamples)+1;
indexB=_impl_-(indexS-1)*wsamples;
****************************************
*/

 indexL=_mult_;
 indexS=floor((_impl_-1)/5)+1;
 indexB=_impl_-(indexS-1)*5;
```

```
run ;

/* Save imputed data */
data d.imputed_synthpops ;
 set synthpops ;
 male=(riagendr=1) ;
run ;

proc sort data=d.imputed_synthpops;
 by indexS indexB indexL;
run ;

/* Estimate of the population mean of lbxtc and its
variance involves 2 steps:
Step 1: Average over IndexB and IndexL for each level
of IndexS */

proc means data=d.imputed_synthpops noprint  mean;
var lbxtc;
by indexS;
output out=step1 mean=lbxtcbar;
run ;

/* Step 2: Compute the mean and variance across the
 S synthetic populations */

proc means data=step1 mean var ;
var lbxtcbar;
run ;

/* Linear Regression analysis using PROC REG with
imputed synthetic populations*/
proc reg data=d.imputed_synthpops;
 by indexS indexB indexL;
 model lbxtc = bmxbmi male indfmpir ;
 ods output parameterestimates=outparms ;
run ;

title "Print Out from Linear Regression" ;
proc print data=outparms ;
run ;

/* prepare combined estimates and variance using
 two data steps*/
proc means data=outparms mean ;
```

```
 var estimate ;
 where variable ='Intercept' ;
 by indexs ;
 output out=step1_0 mean=bobar ;
run ;

proc print data=step1_0 ;
run ;

proc means data=outparms mean ;
 var estimate ;
 where variable ='BMXBMI' ;
 by indexs ;
 output out=step1_1 mean=b1bar ;
run ;

proc print data=step1_1 ;
run ;

proc means data=outparms mean ;
 var estimate ;
 where variable ='male' ;
 by indexs ;
 output out=step1_2 mean=b2bar ;
run ;
proc print data=step1_2 ;
run ;

proc means data=outparms mean ;
 var estimate ;
 where variable ='INDFMPIR' ;
 by indexs ;
 output out=step1_3 mean=b3bar ;
run ;

proc print data=step1_3 ;
run ;

* Merge temp data sets into 1 for combining ;
data step1_all ;
 merge step1_0 step1_1 step1_2 step1_3 ;
 by indexs ;
run ;

proc print data=step1_all ;
```

```
run ;

* use merged data above for final step ;
proc means data=step1_all mean var noprint;
 var bobar b1bar b2bar b3bar;
 output out=step2 mean=intercept bmxbmi male indfmpir
  var=vintercept vbmxbmi vmale vindfmpir;
run ;

proc print data=step2 ;
run ;

/* Combine results for parameter estimates from above */
data combine_linear  ;
 set step2;

 df=5-1 ; *Min(S-1,C-H);
 tvalue=quantile('T',0.975,df);

 /* Create arrays for estimate variance se and
 lower/upper CI */
 array estimate[4] intercept bmxbmi male indfmpir;
 array variance[4] vintercept vbmxbmi vmale vindfmpir;
 array se[4] se_intercept se_bmxbmi se_male se_indfmpir;
 array lower95[4] l95_intercept l95_bmxbmi
 l95_male l95_indfmpir;
 array upper95[4] u95_intercept u95_bmxbmi
 u95_male u95_indfmpir;
 do i=1 to 4;
  se[i]=sqrt((1+1/5)*variance[i]);  * Note that
  denominator must match the number used for
   "csamples" in the code ;
  lower95[i]=estimate[i]-tvalue*se[i];
  upper95[i]=estimate[i]+tvalue*se[i];
 end;
 drop i;
run ;

options nodate nonumber ;
proc print data=combine_linear ;
 title "Combined Estimates, SE, Lower and Upper CI from
 Linear Regression" ;
 var    intercept se_intercept l95_intercept
  u95_intercept
  bmxbmi se_bmxbmi l95_bmxbmi u95_bmxbmi
```

```
          male se_male l95_male u95_male  indfmpir se_indfmpir
           l95_indfmpir u95_indfmpir
           ;
run ;


************************************************************;


/* Logistic Regression analysis using PROC LOGISTIC,
outcome is obese predicted by male, family income to poverty
  and age categories*/

data imputed_synthpops2 ;
 set d.imputed_synthpops ;
 if bmxbmi >=30 then obese = 1 ; else obese=0 ;
run ;

/*predict probability of being obese by gender and age in
 categories and family income to poverty ratio */
proc logistic data=imputed_synthpops2;
 by indexS indexB indexL;
 model obese (event='1') = male indfmpir ag3044 ag4559 ag60 ;
 ods output parameterestimates=outparms_log ;
run ;

proc print data=outparms_log ;
run ;

/* prepare combined estimates and variance using
 two data steps*/
/* Create separate mean by IndexS for each variable */

proc means data=outparms_log mean ;
 var estimate ;
 where variable ='Intercept' ;
 by indexs ;
 output out=step1_0 mean=bobar ;
run ;

proc print data=step1_0 ;
run ;

proc means data=outparms_log mean ;
 var estimate ;
 where variable ='male' ;
 by indexs ;
```

```
 output out=step1_1 mean=b1bar ;
run ;

proc print data=step1_1 ;
run ;

proc means data=outparms_log mean ;
 var estimate ;
 where variable ='INDFMPIR' ;
 by indexs ;
 output out=step1_2 mean=b2bar ;
run ;
proc print data=step1_2 ;
run ;

proc means data=outparms_log mean ;
 var estimate ;
 where variable ='ag3044' ;
 by indexs ;
 output out=step1_3 mean=b3bar ;
run ;

proc print data=step1_3 ;
run ;

proc means data=outparms_log mean ;
 var estimate ;
 where variable ='ag4559' ;
 by indexs ;
 output out=step1_4 mean=b4bar ;
run ;

proc print data=step1_4 ;
run ;

proc means data=outparms_log mean ;
 var estimate ;
 where variable ='ag60' ;
 by indexs ;
 output out=step1_5 mean=b5bar ;
run ;

proc print data=step1_5 ;
run ;
```

```
data step1_all_log ;
merge step1_0 step1_1 step1_2 step1_3 step1_4 step1_5;
by indexs ;
run ;

proc print data=step1_all_log ;
run ;

/* Prepare combined estimates and variance
using two data steps*/
proc means data=step1_all_log mean var ;
 var bobar b1bar b2bar b3bar b4bar b5bar;
 output out=step2_log mean=intercept male indfmpir
 ag3044 ag4559 ag60
 var=vintercept vmale vindfmpir vag3044 vag4559 vag60;
run ;

/*  Combine results for logistic regression */
data combine_log ;
 set step2_log ;
 df=5-1 ;  *Min(S-1,C-H);
 tvalue=quantile('T',0.975,df);

/* Create arrays to the calculations */
array estimate[6] intercept male indfmpir ag3044
 ag4559 ag60;
array variance[6] vintercept vmale vindfmpir vag3044
vag4559 vag60;
array se[6]        se_intercept se_male se_indfmpir
se_ag3044 se_ag4559 se_ag60 ;
array lower95[6]   l95_intercept l95_male l95_indfmpir
 l95_ag3044 l95_ag4559 l95_ag60;
array upper95[6]   u95_intercept u95_male u95_indfmpir
 u95_ag3044 u95_ag4559 u95_ag60;
array or[6]        or_intercept or_male or_indfmpir
 or_ag3044 or_ag4559 or_ag60;
do i=1 to 6;
  se[i]=sqrt((1+1/5)*variance[i]);
  or[i]=exp(estimate[i]);
  lower95[i]=exp(estimate[i]-tvalue*se[i]);
  upper95[i]=exp(estimate[i]+tvalue*se[i]);
end;
 drop i;
run ;
```

```
proc print data=combine_log ;
title "Combined Estimates from Logistic Regression" ;
run ;

</sas>
```

9.2.2   Results

Table 9.1: Estimated Effects of BMI, Gender and Family Income/Poverty Ratio on Total Cholesterol, Linear Regression

	Estimate (SE)	95% CI
Body Mass Index	0.276 (0.064)	(0.083,0.468)
Gender (Male)	-10.682 (0.565)	(-12.251,-9.112)
Ratio Family Income/Poverty	2.798 (0.353)	(1.819,3.777)

Table 9.1 includes parameter estimates, standard errors, and 95% confidence intervals for the linear regression of total cholesterol on gender and family income/poverty ratio. These results account for the complex sample design features, weights, and increased variability due to the imputation process since we are using the "uncomplexed" and weighted data set along with the correct FPBB combining rules.

The results suggest that each predictor is significant at the alpha=0.05 level. For example, a one unit increase in BMI results in an estimated 0.275 increase in total cholesterol, being male results in an estimated reduction of 10.6 points in total cholesterol, as compared to women, and each unit increase in the family income/poverty ratio results in an estimated 2.8 point increase in total cholesterol.

Table 9.2: Estimated Effects of Age, Gender and Family Income/Poverty Ratio on Obesity Status, Logistic Regression

	Odds Ratio	95% CI
Age 30-44	1.586	(1.155, 2.179)
Age 45-59	2.096	(1.605, 2.738)
Age 60+	1.720	(1.236, 2.396)
Male	0.886	(0.733,1.069)
Ratio Family Income/Poverty	0.906	(0.831, 0.987)

Reference Groups are Age 18-29 and Female.

Table 9.2 includes estimated odds ratios and 95% confidence intervals from logistic regression of obesity status (coded 1=obese and 0=not obese) regressed on age, gender, and income/poverty ratio.

Based on Table 9.2, each predictor in this model is statistically significant at the alpha=0.05 level. The results suggest that those in older age groups are significantly more likely to be obese than those age 18-29, males are less likely than females to be obese, and those with higher income/poverty ratios are less likely to be obese, as compared to lower ratios, always holding all else equal.

9.3 Additional Reading

For an overview of complex sample designs, the classic references are Kish (1965) and Cochran (1977). Modern references include Lohr (2009). For analysis of complex survey data, some key references are Kish and Frankel (1974), Rust (1985), Graubard and Korn (1999), Chambers and Skinner (2003), Heeringa et al (2017). The importance of using design variables in the imputation process is discussed in Reiter et al (2006). For more details about the procedure described in this chapter see Zhou, Elliott and Raghunathan (2016a, 2016b, 2016c) and Dong, Elliott and Raghunathan (2014a, 2014b). See also He et al. (2010) for a practical example. These are just a few of many good references in this particular field of research.

9.4 Exercises

1. This exercise repeats Chapter 2., Exercise 1. but requests use of the BBDESIGN command to implement the FPBB approach presented in this chapter rather than the design-based variance estimation approach covered in previous chapters.

   (a) Download the Health and Retirement Survey 2012 data set called **EX_HRS_2012** from the book web site. The analysis goal is to estimate mean BMI in the total population and by gender.

   (b) Repeat the examination of the missing data pattern. How many variables have fully observed and missing data? What is the overall pattern of missing data?

   (c) Use the BBDESIGN command to expand the data set to represent the complex sample design features such as stratification, clusters, and weights. Create 25 implicates in this step.

   (d) Impute any missing data in the expanded data set using the IMPUTE command and request M=1 with 5 iterations. Make sure to use the observed minimum and maximum as imputation bounds for the BMI variable (R11BMI), use a SEED value to allow replication of results, and omit HHID, and PN from the imputation models.

(e) Perform descriptive analysis of mean BMI in the total sample, using the DESCRIBE command without the stratum, cluster, and weight statements. Use the imputed and expanded data set in this step. Make sure to code or calculate the correct variance estimates following the example in this chapter. (Recall that the combining rules for this process are different than for MI analysis). Based on these results, what is the estimated mean BMI (SE) for the population of inference? Why do we omit the design features and weights for the analysis?

2. Repeat Chapter 3, Exercise 3 but again use the BBDESIGN command to implement the FPBB method rather than a design-based method.

(a) Download the data set called **EX_SUBSET_NHANES_0506** from the book web site and if needed, convert to a data set appropriate for use in your software of choice. Note, this data set is restricted to those age 18+ and contains a subset of variables for imputation and analysis, n=5,563. Age has been centered by subtracting mean age in the subpopulation of adults (45.60) from the original age variable.

(b) Use the BBDESIGN command to expand the data set to "uncomplex" or expand the data set.

(c) Examine the extent and pattern of missing data in the expanded data set from part (b). Impute missing data using M=1, a seed value, bounds for the blood pressure variable (your choice), transfer the case ID and age 18+ indicator.

(d) Using the imputed data set from part (c), run a "preliminary" regression model of Diastolic Blood Pressure regressed on gender and centered age. Examine the residual*centered age plot and evaluate the results. What does the plot suggest?

(e) Add a squared age term to the model and repeat step c. Do you see improvement in the residual*centered age plot when the squared term is added?

(f) Run your "final" linear regression and present combined Parameter Estimates, SE, T Tests, and p Values using the correct combining rules. Provide a short paragraph, as for publication, interpreting these results including describing the process used for the BBDESIGN command, the imputation approach, and how the combining rules were implemented.

10

Sensitivity Analysis

10.1 Introduction

So far, all of the analyses assumed ignorable missing data mechanisms and, therefore, no explicit specification of the missing data mechanism was needed. Though this assumption may be reasonable, especially, if covariates with good predictive power for both response indicators and the variables with missing values are available and included in the imputation process. Consider the bivariate example discussed in Chapter 1, where Y_1 is a variable with no missing values and Y_2 has some missing values. Let R_2 be the response indicator variable taking the value 1 for subjects with observed Y_2 and 0 for those with missing Y_2. The missing at random assumption suggests that $Pr(R_2 = 1|Y_1, Y_2, \phi) = Pr(R_2|Y_1, \phi)$ (or more, generally, $Pr(R_2 = 1|Y_1, Y_{2,obs}, \phi)$ where $Y_{2,obs}$ are the observed portion of Y_2). If Y_1 is a good predictor of Y_2 then a considerable fraction of missing information about missing Y_2 can be recovered. Nonetheless, the ignorable missing data mechanism is an unverifiable assumption.

What if that assumption is not true? An explicit mechanism has to be specified which is also external to the observed data. That is, there is no information in the observed data to estimate such a mechanism. For the bivariate example, a Missing Not at Random (MNAR) mechanism can be expressed as $Pr(R_2 = 1|Y_1, Y_2, \phi) = Pr(R_2 = 1|Y_1, Y_{2,obs}, Y_{2,mis}, \phi) = g(Y_1, Y_2, \phi)$. For example, $g(Y_1, Y_2, \phi) = \Phi(\phi_o + \phi_1 Y_1 + \phi_2 Y_2)$ where $\Phi(.)$ is the cumulative normal density function. Obviously, this model cannot be estimated because for every subject with $R_2 = 0$, Y_2 is missing. Specifically, the parameter ϕ_2 cannot be estimated. [Note that, if $\phi_1 = 0$ then ϕ_2 can be estimated under the assumed probit model. The estimates are highly sensitive to the normality assumption.]

Let $f(y_1, y_2|\theta)$ be the joint density function of (Y_1, Y_2). The joint density function of (Y_1, Y_2, R_2) is

$$f(y_1, y_2, r_2|\theta, \phi) = \left\{ g(y_1, y_2, \phi)^{r_2} (1 - g(y_1, y_2, \phi))^{1-r_2} \right\} \times f(y_1, y_2|\theta)$$

163

The predictive distribution of y_2 given (y_1, r_2, ϕ, θ) can be constructed from the above equation (collecting the terms of y_2 and treating all others as constants). Once the missing values of y_2 are imputed, the draws of the parameters θ can be obtained from the complete data posterior density,

$$\pi(\theta|(y_{1i}, y_{2i}), i = 1, 2, \ldots, n) \propto \left\{ \prod_i^n g(y_{1i}, y_{2i}|\theta) \right\} \pi(\theta),$$

where $\pi(\theta)$ is the prior density of θ. Similarly, ϕ can be drawn from its posterior density,

$$\eta(\phi|(y_{1i}, y_{2i}, r_{2i}), i = 1, 2, \ldots, n) \propto$$

$$\left\{ \prod_i g(y_{1i}, y_{2i}, \phi)^{r_{2i}} (1 - g(y_{1i}, y_{2i}, \phi)^{1-r_{2i}} \right\} \eta(\phi)$$

where $\eta(\phi)$ is the prior density of ϕ.

Thus, the Gibbs sampling approach can be applied to create imputations under the non-ignorable model. Draw initial values of missing Y_2 values (say, from an ignorable model) and then draw the parameters, θ and ϕ, from their completed data posterior distributions, and then redraw the missing values from the correct posterior predictive distribution. Continue the process until convergence is achieved for the parameters of interest. Note that some parameters in ϕ (like ϕ_2 in the Probit model) are not estimable and those parameters will have to be drawn from the prior distribution or fixed *a priori*. By varying values of the non-estimable parameter (say, varying values of ϕ_2 in the Probit model), a sensitivity analysis may be performed.

In practice, this is called a selection model approach where the joint distribution $[Y, R]$ is factored into $[Y][R|Y]$. To implement this approach, specialized code has to be developed and is specific to the model specifications g and f. An alternative factorization is through a pattern-mixture model, $[Y, R] = [R][Y|R]$. An advantage of this approach is that it can be implemented by perturbing the imputations obtained under the MAR assumption.

10.2   Pattern-Mixture Model

Let $Pr(R_2 = 1) = \pi$ be the marginal probability of being a respondent. An estimate of π is the proportion of missing values in

Y_2. Let $f(y_2|y_1, R_2 = 1)$ be the conditional density of Y_2 given Y_1 for the respondents. This conditional density can be developed and estimated from the observed data. The conditional density function for the non-respondents is $f(y_2|y_1, R_2 = 0)$, which cannot be estimated from the observed data (just as some parameters in ϕ, in the selection model approach). For example, suppose that $f(y_2|y_1, R_2 = 1, \theta) \sim N(\theta_o + \theta_1 y_1, \sigma_1^2)$. One way to formulate the non-ignorable missing data mechanism is to specify $f(y_2|y_1, R_2 = 0, \theta) \sim N(a(\theta_o + \theta_1 y_1) + b, c^2\sigma_1^2)$ where (a, b, c) define the extent of non-ignorability. The constants a and b define the shift and tilt of the regression line for the non-respondents and c governs the increase or decrease in the residual variance. Sensitivity analysis can be performed by choosing different values of (a, b, c).

This strategy is easy to implement once the multiple imputations have been created under the ignorable missing data mechanism. For every choice of (a, b, c), perturb the imputed values in each imputed data set. Apply the combining rules to obtain inference about the parameters. The sensitivity can be displayed by, say, plotting confidence intervals as a function of (a, b, c).

Suppose that Y_2 is a binary variable and the MAR imputation model is the logistic model $logit Pr(y_2|y_1, \theta) = \theta_o + \theta_1 y_1 = l(y_1)$. The MAR imputation is carried out as follows: Generate an uniform random number u and define the imputed value as 1 if $\log[u/(1 - u)] < l(y_1)$ and 0 otherwise.

Suppose the perturbation proposed for a non-ignorable mechanism is to replace some of the imputed 1's (0's) (under the MAR mechanism) to 0's (1's) with certain probabilities. This may be specified through the specification

$$Pr(y_{2,NMAR} = 0|y_{2,MAR} = 1) = a,$$

and

$$Pr(y_{2,NMAR} = 1|y_{2,MAR} = 0) = b.$$

As before, various choices of (a, b) can used to display sensitivity of inferences to MAR assumption. The next section considers several examples to illustrate this strategy.

10.3   Examples

10.3.1   Bivariate Example: Continuous Variable

This example uses the Opioid Detoxification data set and examines the visual analog score (VAS) day 14 measurement predicted by treatment status (using linear regression). The point of the example is to perturb the day 14 VAS score to examine sensitivity to a MAR assumption. Use of SAS with the XML editor is demonstrated.

The example follows Example 7.5 of Raghunathan (2016) but for simplicity, uses bivariate linear regression rather than a multivariate approach. The day 14 VAS score is perturbed in three ways, once for the treated and not treated groups together and then separately for each treatment group and also for differing levels of perturbation.

The code below imputes missing data using IMPUTE and then perturbs imputed values (assuming MAR) for the VAS day 14 outcome variable. A linear regression model is run for the MAR imputed values and then repeated for models with the outcome perturbed by multiplicative factors of 1.025, 1.05, 1.10, 1.20, and 1.30. SAS PROC REG and PROC MIANALYZE are used to run linear regression and combining of results, though these steps could be performed with REGRESS of IVEWARE or other software tools as well.

```
<sas name="Example 1 Chapter 10">
/* Set libname */

libname d3 'P:\IVEware_and_MI_Applications_Book
\DataSets\Longitudinal Data Set';

proc means data=d3.od_all n nmiss mean min max ;
class treat ;
run ;

/* impute missing data on outcomes, assume MAR */
<impute name="od_imp">
 title "Opioid Detox Impute" ;
 datain d3.od_all;
 dataout odimp all;
 default continuous ;
 categorical female white ;
 transfer instudy usubjid ;
 bounds vas0 (>=0, <=100) vas1 (>= 0, <=100)
 vas2 (>=0, <=100) vas3 (>= 0, <=100) vas4 (>=0, <=100)
  vas5 (>= 0, <=100)
```

```
vas6 (>=0, <=100) vas7 (>= 0, <=100) vas8 (>=0, <=100)
 vas9 (>= 0, <=100)  vas10 (>=0, <=100) vas11 (>= 0, <=100)
vas12 (>=0, <=100) vas13 (>= 0, <=100)  vas14 (>= 0, <=100);
 iterations 5 ;
 multiples 25 ;
 seed 2017 ;
run;
</impute>


/* analyze imputed data sets and do adjustments to
MAR imputed Y values, both groups combined*/
data odimp1;
 set odimp ;
 _Imputation_=_mult_;
 * perturb y on both treatment groups by + 0.025 ;
 vas14_025=vas14*1.025 ;
 vas14_05=vas14*1.05 ;
 vas14_10=vas14*1.10 ;
 vas14_20=vas14*1.20 ;
 vas14_30=vas14*1.30 ;

  * perturb for treat only ;
 tvas14=vas14 ; tvas14_025=vas14 ;   tvas14_05=vas14 ;
  tvas14_10=vas14 ;  tvas14_20=vas14 ;   tvas14_30=vas14 ;
 if treat=1 then do ;
   tvas14_025=vas14*1.025 ;
   tvas14_05=vas14*1.05 ;
   tvas14_10=vas14*1.10 ;
   tvas14_20=vas14*1.20 ;
   tvas14_30=vas14*1.30 ;
 end ;

  * perturb for not treated only ;
 ntvas14=vas14 ; ntvas14_025=vas14 ; ntvas14_05=vas14 ;
  ntvas14_10=vas14 ;  ntvas14_20=vas14 ; ntvas14_30=vas14 ;
 if treat=0 then do ;
    ntvas14_025=vas14*1.025 ;
   ntvas14_05=vas14*1.05 ;
   ntvas14_10=vas14*1.10 ;
   ntvas14_20=vas14*1.20 ;
   ntvas14_30=vas14*1.30 ;
 end ;
 drop _mult_;
run ;
proc sort ; by _imputation_ ; run ;

%macro r (y,title) ;
options nodate nocenter nonumber;
title &title ;
```

```
proc reg data=odimp1 ;
  by _Imputation_;
  model &y=treat ;
  ods output parameterestimates=outparms ;
run;
proc print data=outparms ;
run ;

proc mianalyze parms=outparms ;
 modeleffects intercept treat ;
 ods output parameterestimates=outestmi_&y ;
run;
proc print data=outestmi_&y ;
run ;

%mend ;
/*perturb all treated or not treated*/
%r(vas14,     VAS14 with Mar) ;
%r(vas14_025, VAS14 * 1.025) ;
%r(vas14_05,  VAS14 * 1.05) ;
%r(vas14_10,  VAS14 *  1.10) ;
%r(vas14_20,  VAS14 * 1.20) ;
%R(vas14_30,  VAS14 * 1.30) ;

data all_both ;
 set outestmi_vas14 outestmi_vas14_025 outestmi_vas14_05
 outestmi_vas14_10 outestmi_vas14_20 outestmi_vas14_30 ;
run ;

title "Both Treated and Not Treated" ;
proc print data=all_both ;
 where parm='treat' ;
run ;

/*perturb treated only */
%r(tvas14,     Treated Perturbed VAS14 with Mar) ;
%r(tvas14_025, Treated Perturbed VAS14 * 1.025) ;
%r(tvas14_05,  Treated Perturbed VAS14 * 1.05) ;
%r(tvas14_10,  Treated Perturbed VAS14*1.10) ;
%r(tvas14_20,  Treated Perturbed VAS14*1.20) ;
%R(tvas14_30,  Treated Perturbed VAS14*1.30) ;

data all_treatonly  ;
 set outestmi_tvas14 outestmi_tvas14_025 outestmi_tvas14_05
  outestmi_tvas14_10 outestmi_tvas14_20 outestmi_tvas14_30 ;
run ;

title "Treated Only" ;
proc print data=all_treatonly ;
```

```
    where parm='treat' ;
run ;


/*perturb not treated only */
%r(ntvas14,     Not Treated Perturbed VAS14 with Mar) ;
%r(ntvas14_025, Not Treated Perturbed VAS14 * 1.025) ;
%r(ntvas14_05,  Not Treated Perturbed VAS14 * 1.05) ;
%r(ntvas14_10,  Not Treated Perturbed VAS14*1.10) ;
%r(ntvas14_20,  Not Treated Perturbed VAS14*1.20) ;
%R(ntvas14_30,  Not Treated Perturbed VAS14*1.30) ;

data all_nottreatonly ;
 set outestmi_ntvas14 outestmi_ntvas14_025 outestmi_ntvas14_05
  outestmi_ntvas14_10 outestmi_ntvas14_20 outestmi_ntvas14_30 ;
run ;

title "Not Treated Only";
proc print data=all_nottreatonly ;
 where parm='treat' ;
run ;

</sas>
```

Table 10.1: Estimated Treatment Effect for Day 14 Visual Analog Score (VAS)

Perturbation	Both Groups	Bup-NX Only	Clonidine Only
0.0% (MAR)	-13.441	-13.441	-13.441
2.5%	-13.777	-13.052	-14.167
5.0%	-14.113	-12.663	-14.891
10.0%	-14.786	-11.885	-16.342
20.0%	-16.130	-10.328	-19.243
30.0%	-17.474	-8.771	-22.144

Based on Table 10.1, even with the various perturbed increases in

VAS day 14 scores, the impact of the treatment drug, Bup-NX, alone is more pronounced than the use of the drug Clonidine.

10.3.2   Binary Example

In this example, data from the Primary Cardiac Arrest study is used to demonstrate sensitivity analysis with a binary variable. This file has substantial missing data on the red blood cell membrane measurement of DHA and EPA (REDTOT). As a reminder, the measurement provides information about levels of acids obtained mainly from eating fish and known to be protective against heart disease and arrest. For this example, REDTOT is considered "non-ignorable". The REDTOT variable is first imputed assuming MAR, then dichotomized into a binary outcome, and perturbed using methods appropriate for binary variables. The analytic goal is to use logistic regression to predict the probability of primary cardiac arrest by gender, smoking status, and a binary variable representing lowest quartile of imputed red blood cell membrane DHA/EPA measurements.

The following syntax demonstrates imputation of missing data using IMPUTE followed by creation of a binary version of REDTOT called LOWRED where those in the lowest quartile of the imputed measure are coded 1 and all others set to 2. Perturbed versions of the binary variable LOWRED (with proportions of .05, .10, and .25) for setting LOWRED=1 to 2 and LOWRED=2 to 1 are prepared for subsequent sensitivity analyses. Logistic regression predicting PCA by age, gender, lowest quartile of DHA/EPA acids, and smoker (coded as previous or current smoker=1 and never smoked=2) is executed for the MAR model and for each of the perturbations, using the REGRESS command. Results are combined using the REGRESS command.

```
<sas name="Example 2 C10">

/* Example 2 is for binary outcome using PCA data
 (working data set is test) */

libname pca 'P:\IVEware_and_MI_Applications_Book
\DataSets\PCA
 and Omega 3 Fatty Acids Data';

data pca ;
  set pca.test ;
```

```
    cardiac_arrest=2 ;
    if casecnt=1 then cardiac_arrest=1 ;
    smoker=2 ;
    if smoke in (2,3) then smoker=1 ;
    if smoke eq . then smoker=.;
    * if smoke status equal 2 or 3 then smoker=1 or yes ;
    keep redtot cardiac_arrest smoker age gender studyid ;
run ;

proc means n nmiss mean min max ;
run ;

/* Impute Missing Data For Controls and Cases */
<impute name="impute">
 title "Impute All Missing Data" ;
 datain  pca ;
 dataout imputeall all;
 default continuous ;
 transfer studyid ;
 categorical smoker cardiac_arrest gender;
 iterations 5 ;
 multiples 10;
 seed 666 ;
run ;
</impute>

/* create variables for use in logistic regression */
data imputeall1 ;
 set imputeall ;
 * create low red blood total if in lowest
 quartile of imputed REDTOT ;
 if redtot <=3.8 then lowred=1 ; else lowred=2 ;
  *1=Yes 2=No ;
run ;

/* Logistic Regression Using Imputed Data Sets */
<regress name="Cardiac Arrest regressed on Low Red
 Blood Gender and Smoker">
 title  "MAR Logistic Regression Cardiac Arrest
  is Outcome" ;
 datain imputeall1;
 link logistic;
 categorical gender smoker cardiac_arrest lowred ;
 dependent cardiac_arrest ;
 predictor age gender smoker lowred ;
```

```
run ;
</regress>

%macro rep(p, np1,np2) ;
 * Perturb selected variable (redtot) using value of
  macro variable p ;
 data imputeall&p ;
  set imputeall1 ;
  * create perturbed binary versions of red blood cell
   membrane, set some no on (2) to yes (1) and set some
    yes to no (same value) ;
  lowred&p=lowred ;
  if (lowred=2 and ranuni(5678) <=&np1) then lowred&p=1 ;
   else if (lowred=1 and ranuni(7888) =>&np2)
        then lowred&p=2 ;
run ;
%mend rep ;

%rep(05, .05, .95) ;
%rep(10, .10, .90) ;
%rep(25, .25, .75) ;

<regress name="Logistic Regression with 0.05 Perturb
 on LowRed">
 title  "05 Adjustment for LowRed: Logistic Regression
 Cardiac Arrest is Outcome" ;
 datain imputeall05;
 link logistic;
 categorical gender lowred05 cardiac_arrest smoker ;
 dependent cardiac_arrest ;
 predictor age gender smoker lowred05;
 run ;
</regress>

<regress name="Logistic Regression with 0.10
 Perturb on LowRed">
 title  ".10 Adjustment for LowRed: Logistic Regression
  Cardiac Arrest is Outcome" ;

 datain imputeall10;
 link logistic;
 categorical gender lowred10 cardiac_arrest smoker ;
 dependent cardiac_arrest ;
 predictor age gender smoker lowred10;
 run ;
```

```
</regress>

<regress name="Logistic Regression with 0.25
Perturb on LowRed">
 title  ".25 Adjustment for LowRed Logistic Regression
  Cardiac Arrest is Outcome" ;
 datain imputeall25;
 link logistic;
 categorical gender lowred25 cardiac_arrest smoker ;
 dependent cardiac_arrest ;
 predictor age gender smoker lowred25 ;
 run ;
</regress>
</sas>
```

Table 10.2: Estimated Effect of Low DHA/EPA Measurements on Primary
Cardiac Arrest

Perturbation	Odds Ratio	95% CI
0.0% (MAR)	2.165	(1.957, 2.395)
5.0%	1.867	(1.693, 2.059)
10.0%	1.715	(1.560, 1.886)
25.0%	1.379	(1.260, 1.507)

Based on Table 10.2, even with perturbations of 5%, 10%, and 25%,
having low levels of DHA/EPA significantly increases the incidence
of cardiac arrest, compared to those with higher DHA/EPA levels
while controlling for age, gender, and smoking status.

10.3.3   Complex Example

The example in this section uses the Round 6 European Social
Survey-Russian Federation data set. Refer to Appendix A for a more
complete description of this data set. The primary goal is to examine sensitivity for an index variable representing overall satisfaction
with life and the economy (SATISFIED) in the Russian Federation.
The index is created from the sum of imputed Satisfaction with
the Economy (STFECO, range 0 to 10) and imputed Satisfaction
with Life (STFLIF, range 0 to 10). Therefore, the summed variable
has a range of 0-20 with higher values indicating more satisfaction.
The subsequent linear regression analyses account for the complex
sample design through use of the stratification and cluster variables
(STRATIFY and PSU) and also are weighted with the design weight
(DWEIGHT). The model of interest is overall satisfaction regressed

on gender and age, using a MAR model plus a few models with a perturbed outcome variable.

The following code illustrates imputation of missing data (IM-PUTE), use of SAS data step code to create the outcome index variable of overall satisfaction and perturbation of the outcome variable using selected inflation factors. The example concludes with use of REGRESS to perform MI and weighted and complex sample design adjusted linear regression.

```
<sas name="Example 3 Chapter 10">

libname r 'P:\IVEware_and_MI_Applications_Book\DataSets\ESS
 Russian Federation Data' ;

data ess6_russia ;
 set r.ess6_sub_russia_13nov2015 ;
 keep idno psu dweight stratify stfeco stflife gndr agea ;
run ;
proc means data=ess6_russia nolabels n nmiss mean min max
    std ;
run ;

/*impute missing data on age and 2 satisfaction with life and
 economy variables*/
<impute name="impute_ESS">
 title "Impute All Missing Data" ;
 datain  ess6_russia ;
 dataout imputeess all;
 continuous agea dweight ;
 transfer idno ;
 categorical stflife stfeco stratify psu ;
 iterations 5 ;
 multiples 10;
 seed 8765 ;
 run ;
</impute>

/*prepare satisfaction index variable */
data imputeess1 ;
 set imputeess ;
 agea=round(agea) ;
/* prepare perturbed versions of satisfied*/
 satisfied=sum(of stflife, stfeco) ;
 satisfied10=satisfied*1.10 ;
 satisfied20=satisfied*1.20 ;
 satisfied30=satisfied*1.30 ;
```

```
run ;

<regress name="LinearRegression 1 ESS Russian Fed
Overall Satisfaction with Life Economy">
 title  "ESS Russian Fed Data Satisfaction Regressed
  on Age and Gender" ;
 datain imputeess1 ;
 categorical gndr ;
 stratum stratify ;
 cluster psu ;
 weight dweight ;
 dependent satisfied ;
 predictor agea gndr ;
 run ;
</regress>

<regress name="LinearRegression 2 ESS Russian
Fed Overall Satisfaction with Life Economy">
 title  "1.10 ESS Russian Fed Data Satisfaction
  Regressed on Age and Gender" ;
 datain imputeess1 ;
 categorical gndr ;
 stratum stratify ;
 cluster psu ;
 weight dweight ;
 dependent satisfied10 ;
 predictor agea gndr ;
 run ;
</regress>

<regress name="LinearRegression 3 ESS Russian Fed Overall
 Satisfaction with Life Economy">
 title  "1.20 ESS Russian Fed Data Satisfaction Regressed
  on Age and Gender" ;
 datain imputeess1 ;
 stratum stratify ;
 cluster psu ;
 weight dweight ;
 categorical gndr ;
 dependent satisfied20 ;
 predictor agea gndr ;
 run ;
</regress>

<regress name="LinearRegression 4 ESS Russian Fed Overall
```

```
Satisfaction with Life Economy">
 title  "1.30 ESS Russian Fed Data Satisfaction Regressed
 on Age and Gender" ;
 datain imputeess1 ;
 stratum stratify ;
 cluster psu ;
 weight dweight ;
 categorical gndr ;
 dependent satisfied30 ;
 predictor agea gndr ;
 run ;
 </regress>
</sas>
```

Table 10.3: Estimated Effects of Age and Gender on Overall Satisfaction with Life and Economy

	Age		Gender	
Perturbation	Estimate (SE)	p	Estimate (SE)	p
0.0% (MAR)	-0.033 (0.006)	0.000	-0.088 (0.193)	0.651
10.0%	-0.036 (0.006)	0.000	-0.097 (0.213)	0.651
20.0%	-0.040 (0.007)	0.000	-0.105 (0.232)	0.651
30.0%	-0.043 (0.007)	0.000	-0.114 (0.251)	0.651

Based on results from Table 10.3, being male is non-significant and negatively associated with overall satisfaction with life, as compared to women while a one year increase in age is significantly and negatively related to overall satisfaction, all else held to 0. These results hold for each of the MAR and three perturbed outcomes.

10.4   Additional Reading

The selection model formulation is based on Heckman (1976) where it was used to model the selection of women into the labor force. Lillard, Smith and Welch (1986) present an application of this approach to handle missing income values in the current population survey. Little (1985) and Little and Rubin (1987) show that this approach is highly sensitive to the stated assumptions. Rubin (1977) proposed the simple mixture model framework for a scalar variable based on a normal distribution. Glynn, Laird and Rubin (1986) report on several simulation studies and consider extensions with covariates. Little (1993, 1994, 1995) includs a series of influential articles advocating the pattern mixture model for handling non-ignorable missing data mechanism in a variety of contexts. See Kaciroti and Raghunathan (2014) on the comparison of sensitivity analysis under the two framework, selection and Pattern-Mixture models.

Pregibon (1977), Little (1982), Nordheim (1984), Baker and Laird (1988) and Stasny (1986) are some references that consider nonignorable missing data models for categorical data. The models consider the joint cross-classification of substantive and missing data indicator variables where the parameters that cannot be estimated are either fixed at various values or handled through a prior distribution.

For a practical application see Ratitch and O'Kelly (2011).

10.5 Exercises

(a) Exercise 1 requires replication of the imputation step of Example 1 (of this chapter) and then extends the analysis by adding additional predictors to the linear regression models. The focus is to experiment with expanded models to explore sensitivity analyses of the MAR assumption.

   i. Download the Opioid Detoxification data set from the book web site. Begin by imputing missing data on all variables, as demonstrated in Example 1. Make sure to follow the imputation code and overall logic so that the same bounds and other options are repeated. Use either *IVEware* or similar software for the imputation step.

   ii. Using your software of your choice, create the perturbed outcome variables using the same inflation factors used in Example 1. Why should you use a variety of inflation factors and how does this relate to testing the MAR and NMAR assumptions?

   iii. Using *IVEware* or another software capable of properly combining MI regression results, run the models as in Example 1 but use this expanded model:
   VAS Day 14 (various versions)=Intercept + Age + Female + White + Baseline VAS + Treatment + error

   iv. Prepare a table based on results from (c) and provide general interpretation and a comparison to the results from Example

1. Does the addition of model covariates change your conclusions about the impact of treatment on VAS scores from Day 14 of the study? Why would one treat the VAS-Day 14 Score as "non-ignorable" rather than MAR? Is the MAR assumption used during imputation reasonable and why or why not?

(b) This exercise uses the ESS Russian Federation data set and performs a sensitivity analysis for a binary outcome of high satisfaction with life/economy using two types of models: one is a "naive" approach without weights or design features included and the second is a proper weighted and design-based analysis. The goal of this exercise is two-fold; the first aim is to examine the impact of assuming that the outcome representing high satisfaction is not "ignorable", and the second goal is what impact does ignoring the complex sample features and weights have on analytic conclusions?

   i. Download the ESS Russian Federation data set from the book web site and gather the same set of variables as used in Example 3. Perform exploratory analysis and determine the extent and details of the missing data problem.

   ii. Impute missing data using a software of your choice. Make sure to use a seed value and produce an appropriate number of imputation multiples. What is your rationale for your chosen M and number of iterations? Request imputation diagnostics for one of the imputed variables. Do you see any imputation issues to address? If so, how might you approach the issues?

   iii. Create an index variable which is the sum of the two satisfaction variables, as in Example 3. Make sure that the range of the new variable is 0-20 and that there is no missing data. Next, create a binary indicator of high satisfaction where the top 10% of the continuous variable is set to 1 (high satisfaction) and 2 otherwise ($< 90\%$ on the continuous satisfaction index). Also, create a binary variable called AGECAT coded as 1 if AGEA is between 14 and 30 inclusive, and otherwise coded=2. This variable represents the youngest age group and will be compared to those age 31+.

   iv. Run a logistic regression model as follows: High Satisfaction = youngest age group (AGECAT) and gender (GNDR). Treat each of these variables as categorical. Run the regression once without using weights or the complex sample design variables ("Naive" model) and then repeat the regression but use the weights and design variables in this model ("Weighted and Design-Based" model). Make sure you use a software that is capable of performing MI combining and design-based analyses for this step. These two models are considered MAR analyses.

v. Repeat the steps of part (d) 2 times but alter the binary outcome variable by changing the values of 1 or 2 using the proportions of 0.05 and 0.10 for the re-assignments. Based on the output from the 6 models from parts (d) and (e), prepare a table of Odds Ratios and 95% CI's for the Naive and Weighted/Design-Based models for MAR and the Perturbed Outcomes.

vi. Interpret the results from part (e). Discuss how ignoring the weights and design features impacts significance levels and overall conclusions. Also, consider the results of the sensitivity analysis and treating the outcome as non-ignorable.

11

Odds and Ends

11.1 Imputing Scores

In many situations, a scale is constructed by summing several items and then used as an outcome or predictor variable in a regression analysis. If the items involved in the scale are missing then the scale score is also missing. What are the options for imputing the scale values?

Consider the following approach. Let $X_1, X_2, \ldots, X_p$ be p individual items and the score is the sum $Y = \sum_i^p X_i$. Some times, if the number of missing items is less than k, then the missing items are replaced by the average of the observed items, otherwise the scale is set to missing (that is, if the number of missing items exceeds k). The choice of k is, usually, arbitrary. Furthermore, all the missing items are assumed to be the same for a given individual. Thus, the resulting data set does not resemble any plausible data set from the population. Such ad hoc approaches should be avoided.

Three possible approaches are described depending upon the missing data pattern. The first missing data pattern is called "either all or none". That is, every individual in the sample either responds to all p items or none of the items. In this situation, no partial information is available on subjects with missing values. The scale may then be directly imputed without loss of any efficiency.

Next, consider an arbitrary pattern of missing data among the p items. In this case, it may be better to impute individual items and then construct the sum to create the scale. This allows for imputing the missing values that exploit the correlation among the items and garner the predictive power of observed items to predict the missing items.

This approach can be difficult to implement, if the data set has a lot of items leading to different scale scores. An alternative approach, is to directly impute the missing scores but use the sum of the observed scores as the lower bound in the imputation process. Let Y_{iL} be the sum of the observed items for subject $i = 1, 2, \ldots, n$. Let Y_{max} be the maximum possible value for the scale. The missing value of Y_i may be treated as censored at Y_{iL} and use the "Bounds" feature in *IVEware* to impute the missing value for subject i to be bound between Y_{iL} and Y_{max}. The number of observed items may also be used as a predictor in the imputation regression model for Y.

11.2   Imputation and Analysis Models

Often in practice, more variables may be available than being used in any specific analysis. For example, suppose that the analysis involves a regression model with Y as the dependent variable and $X_1, X_2, \ldots, X_p$ as the covariates. Suppose there are missing values in this analytical variable data set $(Y, X_1, X_2, \ldots, X_p)$ and the missing data mechanism is ignorable. A standard approach (Option 1) is to multiply impute the missing values and then run a regression of observed Y on the observed or imputed $X = (X_1, X_2, \ldots, X_p)$. Note that the imputed values of Y may be ignored in this particular regression analysis because of the following reasoning. Let $R_Y = 1$ if Y is observed and 0 if Y is missing. Let X_{obs} and X_{mis} be the observed and missing portions of X. Under MAR,

$$Pr(Y|R_Y = 1, X_{obs}) = Pr(Y|R_Y = 0, X_{obs}) = Pr(Y|X_{obs})$$

and

$$Pr(Y|R_Y = 1, X_{obs}) \approx \sum_l Pr(Y|R_Y = 1, X_{obs}, X_{mis}^{(l)})/M$$

where $X_{mis}^{(l)}$ are the imputed values of X_{mis}. That is, there is no additional information about the regression of Y on X in the imputed values of Y.

Now suppose that additional variables, $Z = (Z_1, Z_2, \ldots, Z_q)$, are available and are correlated with Y and X. For simplicity, assume that Z is fully observed. The alternative option (Option 2) is to include Z in the imputation model and then fit the analyst regression model with observed Y as the dependent variable and observed or imputed X as predictor variables. For this approach, a number of subtle issues need discussion.

A technical issue is the thought experiment under which the inference procedures are evaluated from the repeated sampling perspective. Under Option 1 (YX imputation and YX analysis), the repeated sampling thought experiment involves sampling of (Y, X) and the missing data mechanism assumes that R, the response indictors, depends only on (Y_{obs}, X_{obs}). The additional variables Z are irrelevant (unrelated to (Y, X) and R) under this thought experiment. Thus, the inclusion of Z in the imputation model is akin to including irrelevant predictors in the regression model, and, hence, inefficient. This option fits well with a typical training in the statistical analysis with no missing data where many analysts select a specific set of variables relevant to their substantive analysis from the large data set and do a specific analysis, ignoring other variables in the data set.

Option 2 (YXZ-imputation, YX-analysis), operates under a different thought experiment where the repeated sampling involves the joint distribution of (Y, X, Z) and R, and depends on (Y_{obs}, X_{obs}, Z) in general, $(Y_{obs}, X_{obs}, Z_{obs})$, if Z also has some missing values. In this situation, the quantity being evaluated is the estimate of a parameter in the YX model.

The choice between Option 1 or Option 2, at least from the technical point of view, depends on the assumptions under which inferential procedures are to be evaluated from the repeated sampling perspective. Our training based on the complete data analysis framework may lead us to consider Option 1 as an extension to analysis with missing data. However, from the perspective of practitioners, Option 2 can lead to more efficient inferences by reducing the uncertainty in the prediction of the missing values and, hence, the expanded thought experiment may have merit.

Now, some caveats. If the additional variables Z are weakly correlated (a simple descriptive measure is R^2 (or Pseudo R^2) from the regression of Z on (Y, X)), then the gain in efficiency may be small. An example is considered later to shed some light on this issue. Imputation model development may become complex with the addition of variables Z as covariates. If a poorly fitting model is used to impute missing values in Y and X, then the estimates may be biased. Additional variables need to be carefully selected and appropriately modeled to reap the benefits to achieve the desired gain in efficiency.

11.2.1 Example

Consider a simple example with two variables, Y, with some missing values and, X, that is fully observed. Assume that the data in Y are missing completely at random (MCAR). The analyst is interested in inferring about the mean $\mu_Y = E(Y)$. Under Option 1, the best estimate of μ_Y is the mean based on r respondents, $\bar{y}_r = \sum_i^r y_i/r$, and its variance estimate is s_{yr}^2/r where $s_{yr}^2 = \sum_i (y_i - \bar{y}_r)^2/(r-1)$, is the sample variance based on the respondent observations.

Suppose that the regression of Y on X is linear. The optimal estimate under Option 2, is the regression estimator,

$$\bar{y}_{lr} = \bar{y}_r + b(\bar{x}_n - \bar{x}_r)$$

where b is the estimated regression coefficient of X in the linear model predicting Y, $\bar{x}_n$ is the full sample mean of X and $\bar{x}_r$ is the mean of X for the respondents in Y. This estimate can be viewed as multiply imputed estimate with infinite number of imputations. The approximate variance of this estimate is

$$var(\bar{y}_{lr}) = \frac{s_{yr}^2}{r}(1 - c^2)$$

where c is the estimated correlation coefficient between Y and X. Thus, the relative efficiency of the regression estimator compared to the respondent mean is

$$RE = \frac{1/var(\bar{y}_{lr})}{1/var(\bar{y}_r)} = (1 - c^2)^{-1}.$$

For example, if $c = 0.5$, the regression estimator is $1/3$ more efficient that the respondent mean but with $c = 0.2$, the regression estimator is only about 4% more efficient than the respondent mean.

Now suppose that the missing data mechanism depends on X (MAR, since X is fully observed) then Option 1 may result in a biased estimate. The analyst needs to incorporate X in the analysis, unless X is uncorrelated with Y. Generally, the inclusion of variables that are related to R_Y (response indicator of Y) but not to the survey measure, Y lead to inefficient estimates. Thus, the choice between Option 1 or Option 2 depends on the kind and amount of information available in the variables not in the substantive model of interest. From a practical point of view, it is important to extract information from all available data even though not all are of interest in a specific analysis model. The goal is to leverage as much information as possible to reduce the uncertainty due to missing values.

11.3   Running Simulations Using *IVEware*

Often, it is useful to run some simulations to evaluate a particular procedure for estimating the parameter of interest. An example of such a situation is whether to use additional covariates Z in the imputation model, as discussed in the previous section. Consider a logistic regression simulation example discussed in Raghunathan (2016). Let D be a binary dependent variable, E, a binary exposure variable and X, a single continuous covariate. The following model is used to generate a complete data set of size $n = 1,000$:

1. $X \sim N(0,1)$,

2. $E|X \sim \text{Bernoulli}(1, \pi(X))$ where $\text{logit}[\pi(X)] = 0.25 + 0.75X$

3. $D|E, X \sim \text{Bernoulli}(1, \theta(X, E))$ where $\text{logit}[\theta(X, E)] = -0.5 + 0.5E + 0.5X$

The model in point 3 is of interest and, in particular, the regression coefficient for E is the target quantity of inference based on a sample from this population.

Now delete some X values based on the following response mechanism,

$$\text{logit}[Pr(R_X = 1|D, E)] = -1 - 0.5D - 0.5E + 3D \times E$$

Figure 1.2 in Raghunathan (2016) generated 500 replicates and provided histograms of the complete data (before deletion) and complete-case (after deletion) estimates to illustrate the potential bias in the complete case analysis. Suppose that an additional variable Z is available with mean 0, variance 1 and with correlation ρ to the variable with missing values, X. This is accomplished by assuming $Z|X \sim N(\rho X, 1 - \rho^2)$.

Now consider multiple imputation analysis of these 500 data sets with and without using the additional variable Z. First, generate 500 data sets without missing values, each of size $n = 1000$, and then set some values of X to be missing. The following portion of the SAS-code accomplishes this task:

```
<sas name="sasoutput">
options ls=80 ps=72 nodate nonotes;
%let n=1000;
%let nsimul=500;
%let rho=0.7;
data one (keep=d e x xobs z simul);
call streaminit(123456);
vz=sqrt(1-&rho*&rho);
do simul=1 to &nsimul;
do i=1 to &n;
/* Generate X and Z */
x=rand('Normal');
z=rand('Normal');
z=&rho*x+vz*z;
/* Generate E */
u1=rand('uniform');
pex=1/(1+exp(-0.25-0.75*x));
e=0;
if u1 le pex then e=1;
/* Generate D */
u2=rand('uniform');
pdex=1/(1+exp(0.5-0.5*e-0.5*x));
d=0;
if u2 le pdex then d=1;
/* Generate Response Indicator */
u3=rand('uniform');
prx=1/(1+exp(1-0.5*d-0.5*e+3*d*e));
xobs=x;
rx=0;
if u3 ge prx then rx=1;
if rx=0 then xobs=.;
output;
end;
end;
```

The above code generates a data set with 50,000 rows and 6 variables: Simulation number (*simul*), D, E, X (before deletion), $Xobs$ (X with missing values) and Z.

Next, *IVEware* is used to impute the missing values in X using the following regression model,

$$X = \alpha_o + \alpha_1 D + \alpha_2 E + \alpha_3 D \times E + \alpha_4 Z + \epsilon$$

where $\epsilon \sim N(0, \sigma^2)$. Setting $\alpha_4 = 0$ corresponds to not using Z in the imputation model (that is, treating Z as a "transfer" variable in *IVEware*). Assume a non-informative prior for α and σ, $Pr(\alpha, \sigma) \propto \sigma^{-1}$. The following code invokes the IMPUTE module in *IVEware* to multiply impute (with $M = 10$) the missing values in X without using Z as a predictor (Option 1 in Section 11.2):

```
/* Create a Copy of the data and generate
imputations under Option 1*/
data temp;
set one;
<impute name="temp2">
datain temp;
dataout tempout all;
default continuous;
transfer simul x z;
interact d*e;
multiples 10;
iterations 1;
by simul;
seed 234789;
run;
</impute>
```

Next, fit a logistic regression model with D as the dependent variable, and E and imputed X (note that $Xobs$ in the output multiply data set called "tempout" has both observed and imputed values), extract the point estimates of the regression coefficient of E and its estimated variance.

```
proc sort data=tempout;
by simul _mult_;
/* Fitting the logistic regression model on
each completed data */
proc logistic data=tempout descending outest=results
 noprint covout;
model d=e xobs;
by simul _mult_;
/* Extract the point estimates */
data est;
set results;
if _TYPE_ ne 'PARMS' then delete;
keep simul _mult_ e;
/* Extract the variance estimates */
data cov;
set results;
if _NAME_ ne 'e' then delete;
```

```
u=e;
keep simul _mult_ u;
/* Create a data set with estimates and its
  completed-data variance */
data anal;
merge est cov;
by simul _mult_;
run;
```

The next step involves calculating the multiple imputation estimate of the regression coefficient of E, its multiple imputation variance estimate, the degrees of freedom and confidence interval. The following code accomplishes these tasks:

```
proc means noprint mean var;
var e u;
output out=res mean=ebarmi ubarmi var=bmi umi;
by simul;
run;

data res;
set res;
drop umi; /* Variance of completed-data variances is not needed */
tmi=ubarmi+(1+1/10)*bmi;
rmi=(1+1/10)*bmi/tmi;
semi=sqrt(tmi);
numi=9/rmi/rmi;
tval=quantile('T',0.975,numi);
lower=ebarmi-tval*semi;
upper=ebarmi+tval*semi;
/* Generate coverage indicator 1: if the interval contains the
true value 0.5 and 0 otherwise */
cov=1;
if lower > 0.5 or upper < 0.5 then cov=0;
/* Length of the confidence interval */
length=2*tval*semi;
run;
```

The final step is to compute the Monte Carlo averages of the point estimates, $\bar{e}_{MI}$, coverage indicator and the length of the confidence interval. Also included is the fraction of missing information.

```
proc means mean var;
var ebarmi cov rmi length;
```

```
run;
</sas>
```

The results from running this program are given below:

The MEANS Procedure

Variable	Mean	Variance
ebarmi	0.5158024	0.0202571
cov	0.9560000	0.0421483
rmi	0.0209743	0.000116975
length	0.5553500	0.000063234

The expected value of $\bar{e}_{MI}$ is 0.52 which is close to the true value 0.5, the actual coverage of the nominal 95% confidence interval is 95.6% and the sampling variance of the estimate is 0.0203. The mean square error is $MSE = (0.5158 - 0.5)^2 + 0.02026 = 0.02051$. The average fraction of missing information is 2.1% which may be compared to 25% of the sample with missing values in X. This suggests that considerable information is recovered through multiple imputation.

To run the simulation under Option 2 (that is, include Z as a predictor in the imputation model), the only change needed to the SAS code is where Z is set as transfer variable in the impute module:

```
transfer simul x;
```

The results from running the simulation under Option 2 are given below:

The MEANS Procedure

Variable	Mean	Variance
ebarmi	0.5121622	0.0202658
cov	0.9500000	0.0475952
rmi	0.0122348	0.000040355
length	0.5527712	0.000057931

The MSE of the estimated regression coefficient when Z is included in the imputation model is 0.02041 (practically the same as without including the Z in the imputation model). However, the average fraction of information is 1.2%, down from 2.1%.

11.4 Congeniality and Multiple Imputations

When assumptions made by the analyst are not "in tune" with the assumptions made by the imputer, multiple imputation inferences can be misleading. Consider the bivariate example in Section 11.2.1. Further assume that the joint distribution of (Y, X) is bivariate normal with means (μ_Y, μ_X), variances (σ_Y^2, σ_X^2), and a correlation coefficient ρ. The goal is to estimate $\theta_c = Pr(Y \geq c)$ for a known constant c. Suppose imputations are developed from draws from the predictive distribution of the missing values in Y conditional on the observed values in (Y, X).

The analyst, however, ignores the model assumptions and uses the empirical estimate,

$$\tilde{\theta}_c^{(l)} = \sum_i^n I_{[y_i^{(l)} \geq c]}/n$$

where $y_i^{(l)}$ is the observed or imputed value for individual i and $I_{[A]}$ is an indicator function taking the value 1 if A is true and 0 otherwise. However, this empirical estimate, though convenient, is not congenial to the assumptions made in the imputation process. What then are the consequences for the analyst in this situation? To explore this further, consider the repeated sampling properties of the multiple imputation estimate,

$$\tilde{\theta}_c^{(MI)} = \sum_1^M \tilde{\theta}_c^{(l)}/M$$

and its variance estimate,

$$\tilde{T}_{MI} = \sum_1^M [\tilde{\theta}_c^{(l)}(1 - \tilde{\theta}_c^{(l)})/n]/M + (1 + 1/M) \sum_1^M (\tilde{\theta}_c^{(l)} - \tilde{\theta}_c^{(MI)})^2/(M - 1).$$

Suppose that 1,000 data sets each of size $n = 250$ are generated from the following model assumptions:

1. $X \sim N(0, 1)$,

2. $Y|X \sim N(\rho X, 1 - \rho^2)$,

3. $\mathrm{logit} Pr(R_Y = 1|X) = -0.5 + 0.25X$ (roughly resulting in 38% missing values in Y).

Assume that the parameter of interest is $\theta_1 = Pr(Y \geq 1)$, (the true value, given that $Y \sim N(0, 1)$, is 0.1587). The three quantities compared through simulations are (1) the bias of $\tilde{\theta}_1^{(MI)}$, (2) Coverage properties of the multiple imputation confidence intervals for θ_1, and (3) the difference between the Monte Carlo variance of the point estimates (approximating the sampling variance of the MI estimates, $Var(\tilde{\theta}_1^{(MI)}))$ and the Monte Carlo average of $\tilde{T}_{MI}$ (approximating the expected value of the variance estimates).

The code given at the end of this section creates a macro to run the simulation for a given value of ρ (input as a macro variable), runs the analysis for three values of $\rho = 0.2, \rho = 0.5$, and $\rho = 0.7$. For $\rho = 0.2$, the expected value of the estimates is $E(\tilde{\theta}_1^{(MI)}) = 0.1575$, the actual coverage of the nominal 95% confidence interval is 99.3%, the sampling variance $Var(\tilde{\theta}_1^{(MI)}) = 0.00036$ and the expected value of the variance is $E(\tilde{T}_{MI}) = 0.00076$. The multiple imputation inferences lead to conservative inferences with over-estimation of the actual sampling variance and over-coverage of the confidence intervals. Similar conservative results are obtained for $\rho = 0.5$ and $\rho = 0.7$. Given the large fraction of missing information, one may want to consider increasing the number of imputations. The same code re-run with $M = 100$ resulted in the confidence interval coverage to be 96.5%, though over-estimation of the sampling variance persists. Generally, when the analyst uses inefficient completed-data estimates relative to the imputation model assumptions, the multiple imputation inferences are conservative.

A serious issue may arise when the imputation model is misspecified. Suppose that the true model in (2) is $Y|X \sim N(\rho X, (1-\rho^2)X^2)$ and the imputer ignores this heteroscedasticity. This type of un-congeniality can be avoided by making sure that the regression models are well fit to the observed data, substantively sensible, and carefully checked by performing model diagnostics. Additionally, the estimation procedures used should be fully efficient to the extent possible to avoid overly conservative inferences.

11.4.1 Example of Impact of Uncongeniality

```
<sas name="uncongen1">
options ls=80 ps=72 nodate nonotes;
/* Create a Macro Environment to run the simulation for different
values of rho, the correlation between Y and X */
%macro uncong(rho);
%let n=250; /* sample size */
%let nsimul=1000; /* Number of Replications */
/* Generate Data Sets */
data one (keep=y yobs x simul);
call streaminit(123456);
vh=sqrt(1-&rho*&rho);
do simul=1 to &nsimul;
do i=1 to &n;
x=rand('Normal');
temp=rand('Normal');
y=&rho*x+vh*temp;
u1=rand('uniform');
prx=1/(1+exp(0.5-0.25*x));
```

```
yobs=y;
ry=0;
if u1 ge prx then ry=1;
if ry=0 then yobs=.;
output;
end;
end;
run;

data temp;
set one;
/* Multiply Impute the Missing Values */
<impute name="uncongen2">
datain temp;
dataout tempout all;
default continuous;
transfer simul y;
multiples 10;
iterations 1;
by simul;
seed 234789;
run;
</impute>

/* Construct completed Data Estimates */
data tempout;
set tempout;
ryc=0;
if yobs >= 1 then ryc=1;
proc sort data=tempout;
by simul _mult_;
run;

proc means noprint mean;
var ryc;
output out=all mean=tildec;
by simul _mult_;
run;

/* Multiple Imputation Analysis */
proc means noprint mean var;
var tildec;
output out=res mean=tildecbar var=bmi;
by simul;
run;
```

```
data res;
set res;
ubarmi=tildecbar*(1-tildecbar)/&n;
tmi=ubarmi+(1+1/10)*bmi;
rmi=(1+1/10)*bmi/tmi;
semi=sqrt(tmi);
numi=9/rmi/rmi;
tval=quantile('T',0.975,numi);
lower=tildecbar-tval*semi;
upper=tildecbar+tval*semi;
truetheta=1-cdf('normal',1);
cov=1;
if lower > truetheta or upper < truetheta then cov=0;
length=2*tval*semi;
run;

/* Output Monte Carlo Means and Variances */
proc means mean var;
var tildecbar tmi cov rmi length;
run;
%mend;
%uncong(0.2);
%uncong(0.5);
%uncong(0.7)
</sas>
```

11.5   Combining Bayesian Inferences

Software packages for performing Bayesian analysis, such as Stan, JAGS, Winbugs, Openbugs, and PROC MCMC etc., have made implementation relatively easy for various model specifications. A common theme in Bayesian analysis is to obtain draws from the posterior distribution of the parameters of interest and use the draws to construct relevant inferences. Thus, one possible approach is use of a Bayesian method for each completed data set and combine the draws across the data sets.

Suppose that a sample of observations $Y = \{y_1, y_2, \ldots, y_n\}$ is generated from the model $f(y|\theta)$ and the prior distribution of θ is $\pi(\theta)$. Also suppose that some values in Y are missing and, as before, $Y = \{Y_{obs}, Y_{mis}\}$. Under the ignorable missing data mechanism, the relevant quantity for constructing the inferences is the posterior distribution of θ, conditional on Y_{obs}, with the density function:

$$\pi(\theta|Y_{obs}) = \int \pi(\theta|Y_{obs}, Y_{mis})Pr(Y_{mis}|Y_{obs})dY_{mis},$$

where

$$\pi(\theta|Y = \{Y_{obs}, Y_{mis}\}) \propto \pi(\theta) \prod_i^n f(y_i|\theta)$$

is the complete data posterior density function and $Pr(Y_{mis}|Y_{obs})$ is the predictive distribution of the missing set of observations. If $Y_{mis}^{(l)}, l = 1, 2, \ldots, M$ are the imputations drawn from the predictive distribution, $Pr(Y_{mis}|Y_{obs})$, then the above integral is approximately equal to

$$\frac{1}{M} \sum_{l=1}^M \pi(\theta|Y_{obs}, Y_{mis}^{(l)}),$$

a mixture of M density functions with mixing probability, $1/M$. Conceptually, the following procedure draws from this mixture distribution : Randomly select a number between 1 and M, say, j, and then obtain a draw of θ from $\pi(\theta|Y_{obs}, Y_{mis}^{(j)})$. Repeating this process, say R times, obtain approximately independent draws from the mixture distribution given above.

Computationally, of course, it makes more sense to perform the Bayesian analysis on each completed data set to obtain R draws of the parameters from each completed data set. Suppose that these draws are arranged in a $R \times M$ matrix, then all RM draws can be combined to form a single inference as suggested by Gelman et al (2004). For example, the empirical percentiles (useful for constructing Bayesian credible intervals) of the posterior distribution of θ can be calculated using ordered RM values. An alternative is to use the draws from each completed data posterior distributions to compute the completed data posterior mean and variance and then combine them using the standard formula given in Section 1.13. This approach may not work well for all estimands, as shown by Zhou and Reiter (2010). Also, they show that the suggestion by Gelman et al (2004) may work well only when M is large.

Now consider an elaboration of the approach suggested by Gelman et al (2004). The conceptual procedure is equivalent to drawing one element at random from each row of this matrix to generate R draws from the mixture distribution. Let θ_{rl} denote the rth draw from the completed data posterior distribution with the density $\pi(\theta|Y_{obs}, Y_{mis}^{(l)}), r = 1, 2, \ldots, R; l = 1, 2, \ldots, M$. Let θ_r^* denote a random draw from the row $r = 1, 2, \ldots, R$. The posterior mean is then approximated by $\bar{\theta}^* = \sum_r \theta_r^*/R$ and the posterior variance by $T^* = \sum_r (\theta_r^* - \bar{\theta}^*)^2/(R-1)$.

Clearly, $\bar{\theta}^*$ and T^* are not using all other generated values and this approach is wasteful. Suppose that the process of drawing one random element from each row and then computing the posterior mean and variance is repeated infinitely many times and then both $\bar{\theta}^*$ and T^* are averaged. Thus, with respect to this sampling process, in expectation, $E(\theta_r^*) = \bar{\theta}_{r+} = \sum_l \theta_{rl}/M$ and $E(\bar{\theta}^*) = \sum_r \bar{\theta}_{r+}/R = \bar{\theta}_{++}$, the overall mean of all RM draws, $\theta_{++} = \sum_r \sum_l \theta_{rl}/(RM)$.

Next, consider the expected value of T^*. Writing $(R-1)T^* = \sum_r \theta_r^{*2} - R\bar{\theta}^{*2}$, it follows that

$$(R-1)E(T^*) = \sum_r (Var(\theta_r^*) + \bar{\theta}_{r+}^2) - R(Var(\bar{\theta}^*) + \bar{\theta}_{++}^2).$$

The variance of θ_r^*, with respect to the sampling process is $(1 - 1/M)v_r$ where $v_r = \sum_l (\theta_{rl} - \bar{\theta}_{r+})^2/(M-1)$ (based on the Simple Random Sampling of 1 element from the population of M elements). Due to independent sampling across the rows, $Var(\bar{\theta}^*) = \sum_r (1 - 1/M)v_r/R^2$. Substitution and simplification produces:

$$(R-1)E(T^*) = (1 - 1/R)(1 - 1/M)\sum_r v_r + \sum_r (\bar{\theta}_{r+} - \bar{\theta}_{++})^2$$

or

$$T_{MI} = E(T^*) = (1 - 1/M)\bar{V}_R + B_R$$

where $B_R = \sum_r (\bar{\theta}_{r+} - \bar{\theta}_{++})^2/(R-1)$ (the Between-Row Mean Square) and $\bar{V}_R = \sum_r v_r/R$ (the Within-Row Mean Square). Note that when R and M are large enough to ignore the term $(1 - 1/R)(1 - 1/M)$, T_{MI} is nearly equal the overall mean square $\sum_r \sum_l (\theta_{rl} - \bar{\theta}_{++})^2/(RM-1)$.

Also note that, $U_l = \sum_r (\theta_{rl} - \bar{\theta}_{+l})^2/(R-1)$ where $\bar{\theta}_{+l} = \sum_r \theta_{rl}/R$ is completed-data posterior variance based on the completed data $l = 1, 2, \ldots, M$. Define $\bar{U}_{MI} = \sum_l U_l/M$ (Within-Column Mean Square). The fraction of missing information is $r_{MI} = (T_{MI} - \bar{U}_{MI})/T_{MI} = 1 - \bar{U}_{MI}/T_{MI}$ and the degrees of freedom for the t distribution is $\nu_{MI} = (M-1)/r_{MI}^2$.

Consider the problem discussed in the previous section where the goal is to infer about $\theta_c = Pr(Y \geq c)$. It is fairly easy to draw from the posterior distribution of θ_c from each completed data as follows:

1. Draw a chi-square random variate u with $n-1$ degrees of freedom and define $\sigma_Y^{*2} = (n-1)s_{yn}^{(l)2}/u$

2. Draw a standard normal deviate z and define $\mu_Y^* = \bar{y}_n^{(l)} + \sigma_Y^* z/\sqrt{n}$

3. Compute $\theta_c^* = 1 - \Phi[(c - \mu_Y^*)/\sigma_Y^*]$

4. Repeat Steps (1)-(3), a total of R times, yielding $\theta_{c,r}^*, r = 1, 2, \ldots, R$

5. Repeat Steps (1) to (4) for each completed data set $l = 1, 2, \ldots, M$ to generate a data set with RM values of θ_c.

Next, a series of computations of the means and variances based on these draws is needed to construct inference about θ. The code at the end of this section runs a simulation to evaluate the repeated sampling properties of the estimate

$\bar{\theta}_{MI} = \bar{\theta}_{++}$ and the coverage properties of the nominal 95% confidence interval for θ_1. The setup is exactly the same as described in Section 11.4. The sample size is $n = 1000$, the correlation coefficient between Y and X is $\rho = 0.9$, the number of imputation is $M = 5$, the number of draws from the posterior distribution is $R = 100$ and the number of replications or simulated data sets is 500.

Results from this simulation are $E(\bar{\theta}_{MI}) = 0.1587$, $T_{MI} = Var(\bar{\theta}_{MI}) = 0.000087$, the coverage of nominal 95% confidence interval is 96.2% and $E(T_{MI}) = 0.000095$. The same code/setup was run to evaluate the empirical estimate $\tilde{\theta}_{MI}$ (discussed in the previous section) which resulted in $Var(\tilde{\theta}^{(MI)}) = 0.000118$. The Bayes estimate is almost 36% more efficient than the empirical estimate with a slightly conservative interval estimates.

11.5.1 Example of Combining Bayesian Inferences

```
<sas name="bayesanal1">
options ls=80 ps=72 nodate nonotes;
%let n=1000; /*Sample Size */
%let nsimul=500; /* Number of Replications */
%let rho=0.9; /* Assumed Correlation between Y and X */
%let capr=100;
%let nmult=5;
/* Number of draws from the posterior distribution */
/* Generate Data sets */
data one (keep=y yobs x simul);
call streaminit(123456);
vh=sqrt(1-&rho*&rho);
do simul=1 to &nsimul;
do i=1 to &n;
x=rand('Normal');
temp=rand('Normal');
y=&rho*x+vh*temp;
u1=rand('uniform');
prx=1/(1+exp(0.5-0.25*x));
yobs=y;
ry=0;
if u1 ge prx then ry=1;
if ry=0 then yobs=.;
output;
end;
end;
run;

/* Multiply Impute the Missing Values */
data temp;
```

```
set one;
<impute name="bayesanal2">
datain temp;
dataout tempout all;
default continuous;
transfer simul y;
multiples 5;
iterations 1;
by simul;
seed 234789;
run;
</impute>
/* Compute the mean and variance of imputed Y */
proc sort data=tempout;
by simul _mult_;
run;

proc means noprint mean var;
var yobs;
output out=all mean=ybarnl var=s2ynl;
by simul _mult_;
run;

/* Macro to implement draws from the posterior distribution
for each simulated imputed data set */
%macro bayesanal;
%do i=1 %to &nsimul;
%do j=1 %to &nmult;
data temp;
set all;
where simul=&i and _mult_=&j;
%do r=1 %to &capr;
runnumber=&r;
uu=rand('chisquare',&n-1);
sigstar=sqrt((&n-1)*s2ynl/uu);
zz=rand('normal');
mustar=ybarnl+sigstar*zz/sqrt(&n);
theta1star=1-cdf('normal',1,mustar,sigstar);
output;
%end;
/* Append the Data for each completed data and simulated data */
proc append base=empty data=temp;
%end;
%end;
%mend;
```

```
/* Run the Macro */
%bayesanal;
/* Compute UbarMI */
proc sort data=empty;
by simul _mult_;
run;
proc means noprint var;
var theta1star;
by simul _mult_;
output out=new  var=ul;
proc sort data=new;
by simul;
proc means noprint mean;
var ul;
output out=ubar mean=ubarmi;
by simul;
/* Compute VbarR, BR etc */
proc sort data=empty;
by simul runnumber;
proc means noprint mean var;
var theta1star;
by simul runnumber;
output out=new mean=thetabarr var=srsquare;
proc means noprint mean var;
var thetabarr srsquare;
output out=new2 mean=postmean sb2 var=sa2 junk;
by simul;
run;

/* Merge results for each simulated data and construct
   MI inferences */
data result;
merge ubar new2;
by simul;
tmi=sa2+sb2*(1-1/&nmult);
semi=sqrt(tmi);
rmi=1-ubarmi/tmi;
nu=&nmult/rmi/rmi;
tval=quantile('t',0.975,nu);
lower=postmean-tval*semi;
upper=postmean+tval*semi;
truetheta=1-cdf('normal',1);
cov=1;
if lower > truetheta or upper < truetheta then cov=0;
run;
```

```
proc means mean var;
var postmean tmi cov sa2 sb2;
run;
</sas>
```

11.6   Imputing Interactions

When building a substantive analysis regression model, nonlinear terms or interactions between two or more variables may enter as predictors. If any of the variables have missing values then these derived variables (nonlinear functions or products) are also missing. How should the missing values in these derived variables be handled? One possibility is to add the nonlinear and interaction terms as "new" variables in the data set and impute them just like any other variable. This strategy can create inconsistent values in the imputed data set. For example, suppose that the model involves X_1 and $X_2 = X_1^2$. Whenever X_1 is missing (observed) then X_2 is also missing (observed) and X_2 and X_1 has deterministic relationship. By treating them as two different variables in the imputation process, there is no guarantee that this deterministic relationship will be satisfied. Despite this problem, this approach (called the "Just Another Variable" approach) has been evaluated using simulated data sets and seems to provide reasonable point and interval estimates of the regression coefficients in the analysis regression model.

An alternative approach is to derive the predictive regression models for the missing values that is in tune with the specific analysis model. This problem was considered in Section 1.10.2 and used the Gibbs sampling approach to obtain multiple imputation values/data sets. Given that it is complicated to implement this approach, Section 1.10.1 considered the SRMI framework where the regression models included interaction terms. The goal is to assess whether careful modeling and the inclusion of interaction terms as described in Section 1.10.1 can result in valid estimates of the target parameters of interest.

11.6.1   Simulation Study

To explore this further, consider the following model assumptions:

1. $X_1 \sim N(2,1)$,
2. $X_2|X_1 \sim N(1 + 0.5X_1, 0.86^2)$,
3. $Y|X_1, X_2 \sim N(X_1 + X_2 + X_1X_2, 5.5^2)$

This setup roughly results in R^2 for the model in (3) (the ultimate model of interest) to be approximately 0.5 and the interaction term to be substantively important. Both X_1 and X_2 have the same mean $(E(X_1) = E(X_2) = 2)$ and almost the same variance $(Var(X_1) = 1, Var(X_2) \approx 1)$. The correlation between X_1 and X_2 is approximately 0.5. This setup is similar to that considered by Bartlett et al (2015).

The variable Y is fully observed, both X_1 and X_2 have missing values and the missing data mechanism depends on Y, with 70% complete (that is, no missing values) and the remaining with either X_1 missing or X_2 missing (see the code provided at the end of the section for more details). The sample size is fixed at $n = 1,000$. The first generated data was analysed to develop imputation models for X_1 and X_2 and these were:

M1. $X_1|X_2, Y \sim N(\alpha_o + \alpha_1 X_2 + \alpha_2 Y + \alpha_3 X_2 Y + \alpha_4 X_2^2 + \alpha_5 Y X_2^2, \sigma_1^2)$,

M2. $X_2|X_1, Y \sim N(\beta_o + \beta_1 X_1 + \beta_2 Y + \beta_3 X_1 Y + \beta_4 X_1^2, +\beta_5 Y X_1^2, \sigma_2^2)$

These two models are consistent with the comparison of the joint model and SRMI approaches discussed in Section 1.10.3. Higher order terms (involving X_2 in (1) and involving X_1 in (2)) do improve the model fit but not substantially.

The analysis model of interest is

$$Y = \gamma_o + \gamma_1 X_1 + \gamma_2 X_2 + \gamma_3 X_1 X_2 + \epsilon,$$

where $E(\epsilon|X_1, X_2) = 0$ and $Var(\epsilon|X_1, X_2) = \sigma^2$. Note that the true values of the regression coefficients are $\gamma_o = 0, \gamma_1 = \gamma_2 = \gamma_3 = 1$. Table 11.1 summarizes the results for the three regression coefficients when the imputations are created under the model assumptions (M1) and (M2). The simulation study involves 250 replications with each multiply imputed with $M = 100$. Also, the results are provided for the default models,

D1. $X_1|Y, X_2 \sim N(\alpha_o + \alpha_1 Y + \alpha_2 X_2, \sigma_1^2)$

D2. $X_2|Y, X_1 \sim N(\beta_o + \beta_1 Y + \beta_2 X_1^2, \sigma_2^2)$

When the imputations are created under Models M1 and M2, the estimated regression coefficients are close to their true values (1 for all three coefficients). It can be checked that estimates get closer to 1 when additional terms (such as $Y X_1^3$ and X_1^3 in Model M1 and $Y X_2^3$ and X_2^3 in Model M2) are added in the imputation model. Thus, better imputation models may be developed by including information about the analysis model of interest. Note that default models (not supported by the regression analysis) are misspecified and lead to biased inferences. Also, note that the confidence intervals under (M1,M2) models are conservative, perhaps, a consequence of the imputation model using many parameters and thus increasing uncertainty in the imputed values.

11.6.2   Code for Simulation Study

```
<sas name="interact1">
options ls=80 ps=72 nonotes;
```

Table 11.1: Simulation results using SRMI when the analysis model involves interactions

Description	γ_1	γ_2	γ_3
Imputation Under Models M1 and M2			
Average of Estimates	1.0967	1.0734	0.9581
Variance of Estimates	0.1144	0.1159	0.0228
Average of MI Variances	0.1713	0.1648	0.0292
Coverage (95% nominal)	99.6	98.4	96.8
Imputation Under Default Models D1 and D2			
Average of Estimates	1.3248	1.2447	0.8593
Variance of Estimates	0.1168	0.1170	0.0231
Average of MI Variances	0.1681	0.1625	0.0289
Coverage (95% nominal)	91.6	94.4	88.0

```
/* Generate nsimum data sets each of size n with missing values */
%let nsimul=250;
%let n=1000;
data orig(keep=simul y x1obs x2obs);
call streaminit(123456);
do simul =1 to &nsimul;
do i=1 to &n;
x1=2+rand('normal');
x2=1+0.5*x1+0.86*rand('normal');
y= x1+x2+x1*x2+5.5*rand('normal');
u1=rand('uniform');
u2=rand('uniform');
py=1/(1+exp(-1+0.3*y));
rx1=0;
if u1 >= py then rx1=1;
rx2=0;
if u2 >=py then rx2=1;
u3=rand('uniform'); rx22=0;
if rx1=0 and rx2=0 and u3<=0.75 then rx22=1;
if rx22=1 then rx2=1;
if rx22=0 then rx1=1;
x1obs=x1;
x2obs=x2;
if rx1=0 then x1obs=.;
if rx2=0 then x2obs=.;
output;
end;
end;
```

```
run;

/* Multiply impute the missing values in each simulated data */
<impute name="interact2">
datain one;
dataout oneout all;
default continuous;
interact x2obs*x2obs x2obs*y x2obs*x2obs*y
 x1obs*x1obs x1obs*y x1obs*x1obs*y;
multiples 100;
iterations 10;
seed 2005;
by simul;
run;
</impute>

/* Run the Substantive model on each simulated and imputed data */
data all;
set oneout;
x1x2=x1obs*x2obs;
proc sort;
by simul _mult_;
proc reg noprint outest=results covout;
model y=x1obs x2obs x1x2;
by simul _mult_;
run;

/* Extract the Point Estimates and their variances */
data est;
set results;
if _TYPE_ ne 'PARMS' then delete;
keep simul _mult_ x1obs x2obs x1x2;
data cov1;
set results;
if _NAME_ ne 'x1obs' then delete;
u1=x1obs;
keep simul _mult_ u1;
run;

data cov2;
set results;
if _NAME_ ne 'x2obs' then delete;
u2=x2obs;
keep simul _mult_ u2;
data cov3;
```

```
set results;
if _NAME_ ne 'x1x2' then delete;
u3=x1x2;
keep simul _mult_ u3;
run;

data anal;
merge est cov1 cov2 cov3;
by simul _mult_;
run;

/* Combine the Point eatimates and construct
 multiple imputation inferences */
proc means noprint mean var;
var x1obs x2obs x1x2 u1 u2 u3;
output out=res
mean=ebar1 ebar2 ebar3 ubar1 ubar2 ubar3
var=b1 b2 b3 j1 j2 j3;
by simul;
run;

data res;
set res;
t1=ubar1+(1+1/100)*b1;
t2=ubar2+(1+1/100)*b2;
t3=ubar3+(1+1/100)*b3;
r1=(1+1/100)*b1/t1;
r2=(1+1/100)*b2/t2;
r3=(1+1/100)*b3/t3;
se1=sqrt(t1);
nu1=99/r1/r1;
tval1=quantile('T',0.975,nu1);
lower1=ebar1-tval1*se1;
upper1=ebar1+tval1*se1;
cov1=1;
if lower1 > 1 or upper1 < 1 then cov1=0;
length1=2*tval1*se1;
se2=sqrt(t2);
nu2=99/r2/r2;
tval2=quantile('T',0.975,nu2);
lower2=ebar2-tval2*se2;
upper2=ebar2+tval2*se2;
cov2=1;
if lower2 > 1 or upper2 < 1 then cov2=0;
length2=2*tval2*se2;
```

```
se3=sqrt(t3);
nu3=99/r3/r3;
tval3=quantile('T',0.975,nu3);
lower3=ebar3-tval3*se3;
upper3=ebar3+tval3*se3;
cov3=1;
if lower3 > 1 or upper3 < 1 then cov3=0;
length3=2*tval3*se3;
run;


/* Monte Carlo averages and variances of point estimates,
variance estimates and coverage */
proc means mean var;
var ebar1 ebar2 ebar3 t1 t2 t3 cov1 cov2 cov3;
run;
</sas>
```

11.7 Final Thoughts

Multiple Imputation (MI) is a versatile and practical tool for performing the analysis of incomplete data. Like other approaches such as weighting methods, maximum likelihood, and fully Bayesian analyses based on observed data, MI also involves assumptions and in fact, there are no assumption free approaches for the analysis of incomplete data. All approaches involve assumptions about the mechanism that resulted in the missing values and the models for the analysis variables of interest. MI further parses the assumptions by possibly considering two different models, a model for imputation and a model for the analysis. This allows for the possibility of using a richer set of variables (than those included in the analysis model) to improve the imputation process. The goal, after all, is to obtain inferences about the parameters in the analysis model while reducing the uncertainty about the missing values.

An imputer has to make a careful choice of variables which are predictive, first and foremost, of the variables with missing values, and, if possible, related to the missingness in the data. Exploratory analysis, model building and checking, and diagnostic tools should be used to develop good fitting regression models for predicting variables with missing values. This is akin to a careful analysis that one usually does (or should do) when developing models for scientific inference. Sometimes, there is a tendency to "automate" the imputation process (or there is a desire for automation). Though this desire is understandable, it can lead to problems. No scientific modeling can be fully automated and choosing default options in software packages (which are necessary for compiling the code, are somewhat simplistic, and make routine or standard assumptions) can be quite dangerous. Software packages are just tools and require a careful and deliberate user to make the best use of them.

Some automation is possible, if non-parametric regression models such as Classification and Regression Trees (CART) or Propensity score/Predictive Mean Matching hot-deck (as discussed in Chapter 1) are used for all variables in the SRMI setup. Such procedures, however, may not be efficient and may only be useful when the sample size is large and the fraction of missing information is relatively small.

In practice and due to its flexibility, sequential regression multiple (or multi-variate) imputation (SRMI) is often chosen for handling complex data sets. But, by no means is this the only or the best method for all possible scenarios. This approach is easy to implement as it involves as many regression models when there are a number of variables with missing values under consideration. In addition, it allows for the use of auxiliary variables as predictors, consistencies in values across variables are accommodated in the imputation process, and many software packages have implemented this approach.

Possible incompatibility of the regression models with a joint distribution is an important technical or theoretical issue. That is, the specifications $f(x|y, z)$ and $f(y|x, z)$ for the two variables, (x, y) with missing values and z with no missing values may not correspond to any joint bivariate distribution $f(x, y|z)$. Hence, the convergence properties of the iterative algorithm underpinning the SRMI is not known. Though several conditions have been developed for convergence and numerous simulation studies have shown that the approach results in valid inferences, if the models are not a good fit for the data, the theoretical problem is still an unresolved issue. The development of a good fitting model is important to be in tune with the simulation studies in the literature.

The choice of the particular software, *IVEware* (as an add-on to SAS), is to provide a common thread across various techniques discussed in this book. The same code will work with other packages with minor modifications. Numerous other excellent software packages are available and can be used to perform the same analyses, though not all features in *IVEware* are available in other software packages. Of course, the converse, not all features available in other software packages are available in *IVEware*, is also true. The choice for this book, perhaps, is more for the convenience of the authors, as *IVEware* was built by the authors!

11.8   Additional Reading

Fay (1992) first pointed out the problem with multiple imputation variance estimates under uncongeniality but the term un-congeniality was coined by

Meng (1994) and also provided a conceptual framework for discussing the validity of inferences when the imputation and analysis models differ. A series of papers addressed this issue in great detail and as a result, alternative combining rules have emerged. See, for example, Robins and Wang (2000) and Kim et al (2006). A recent article by Xie and Meng (2017), has been discussed by several authors and a rejoinder may be useful in understanding the issues.

Imputation of interactions has received considerable attention. See for example, Kim, Sugar, and Belin (2015), Von Hippel(2009), Jonathan et al (2015) and Seaman et al (2012).

If one does not have a favorite Bayesian software in place, Stan (Stan Development Team (2016)) is recommended because of its flexibility and expansiveness of model types that can be fit. Stan can run under the *R* package (*RStan*) and therefore, can be integrated with *IVEware*. Another option for SAS users is PROC MCMC though this procedure is not as versatile as the Stan package.

Finally, this book uses multiple imputation as the primary method for incorporating uncertainty due to imputations. There are alternative approaches for drawing correct inferences from the imputed data. Some references include Rao (1996), Rao and Shao (1992), Kim and Shao (2014) and Yang and Kim (2016).

11.9 Exercises

1. **Project**. Consider two variables (Y, X) with X fully observed and normally distributed with mean 0 and variance 1. The variable Y has some missing values with a response mechanism, $\text{logit} \, Pr(R_Y = 1|X) = \alpha_o + \alpha_1 X$. Fix the sample size at $n = 500$. Choose α_o and α_1 to yield 30%, 50% or 70% missing values in Y.

   (a) Generate values of Y from the model $Y|X \sim N(\beta_o + \beta_1 X, \sigma^2)$. The goal is to infer about the mean μ_Y, standard deviation, σ_Y and the regression coefficient σ_{XY}/σ_X^2. Perform imputation under this assumed model and construct multiple imputation inferences for these parameters. Evaluate the repeated sampling properties, bias, variance and confidence coverage by replicating the process $R = 1,000$ times.

   (b) For the same setup as in (a), perform multiple imputation using the non-parametric procedure (described as ABB in *IVEware*), evaluate the repeated sampling properties and compare them those obtained in (a).

   (c) Generate values of Y from the model $Y = \beta_o + \beta_1 X + \epsilon$ where $\epsilon = \sigma(u - 3)/\sqrt{6}$ where u has a chi-square distribution with 3 degrees of freedom. Perform imputations under normal, Tukey's *GH* and ABB procedures (see *IVEware* user guide). Comment on the robustness of imputations performed assuming normality for each of the three parameter estimates.

(d) Generate values of Y from the model $Y|X \sim N(\beta_o + \beta_1 X, \sigma^2 X^2)$. Perform imputations assuming $Y|X \sim N(\beta_o + \beta_1 X, \sigma^2)$ and also using the ABB procedure.Comment on the robustness of imputations performed assuming normality for each of the three parameter estimates.

2. **Project.** Let $X \sim N(0, 1)$ and $Y|X \sim N(\beta_o + \beta_1 X, \sigma^2 X^2)$. Suppose that X is fully observed and Y is missing with the response mechanism, $\text{logit}[Pr(R_Y = 1|X) = \alpha_o + \alpha_1 X$. The goal is to multiply impute the missing values in Y. Define $Z = Y/X$ and set Z to missing whenever Y is missing. Let $V = 1/X$. Use the normal regression procedure to impute the missing values in Z just using the data set with (Z, V) (that is, treat Y and X as transfer variables in *IVEware*), and then define the imputed values of Y as Z/V. Conduct a simulation study to evaluate the multiple imputation estimate of the mean μ_Y and standard deviation σ_Y.

3. As an extension of Problem (2), consider $Y|X \sim N(\beta_o + \beta_1 X, \sigma^2 X^{2\nu})$ where ν is unknown. Suggest an approach for estimating ν and then extend the method discussed in Problem (2). Extend the simulation study in Problem (2) for various choices of ν.

4. Modify the code for the simulation study described in Section 11.6 to include $(X_1^3, X_2^3, X_1^3 Y, X_2^3 Y)$ and then $(X_1^4, X_2^4, X_1^4 Y, X_2^4 Y)$. Do the results improve in terms of reducing the bias in the estimated regression coefficients?

5. Modify the simulation study in Section 11.6 by using the ABB procedure in *IVEware* for imputing the missing values in X_1 and X_2, instead of normal linear regression model. Compare the results with those provided in Section 11.6 and also with those obtained as solutions to Problem (4).

A

Overview of Data Sets

Numerous data sets are used in the text book and these are drawn from a variety of sources such as observational studies, clinical trials, longitudinal studies, and data derived from complex sample designs. For each data set, a general description, contents listing, and basic descriptives of key variables used in applications are presented in this appendix. The selected data sets contain interesting analysis variables along with weights and complex sample design related variables, if applicable. Some variables are fully observed while others have item missing data which requires imputation. In addition to core data sets, a variety of data for use with Chapter Exercises are available from the book web site but not fully documented here due to space considerations.

A.1 St. Louis Risk Research Project

The St. Louis Risk Research Project (SLRRP) was an observational study to assess the effects of parental psychological disorders on various aspects of child development. In a preliminary cross-sectional study, data were collected on 69 families having two children each. Each family was classified into three risk groups for parental psychological disorders. The children were classified into two groups according to the number of adverse psychological symptoms they exhibited. Standardized reading and verbal comprehension scores were also collected for the children. Each family is thus described by four continuous and three categorical variables. Because of its mixture of continuous and categorical variables with missing values, the SLRRP data set has become a classic data set for evaluating imputation methods for mixed data types. See Little and Schluchter (1985), Schafer (1997), Little and Rubin (2002), Liu and Rubin (1998), Raghunathan et al. (2001) and Raghunathan (2016). The key variables and descriptive statistics are listed in Tables A.1 and A.2.

Table A.1: Variables in St.Louis Risk Research Project

Variable Name	Label
FAMID	Family ID
G	Parental Risk Group
R1	Reading Score Child 1
R2	Reading Score Child 2
S1	Symptoms Child 1
S2	Symptoms Child 2
V1	Verbal Score Child 1
V2	Verbal Score Child 2

Table A.2: Descriptive Statistics for the SLRRP Data

Variable	N	N Missing	Mean	Standard Deviation
G	69	0	1.87	0.80
R1	48	21	108.10	16.84
R2	53	16	104.70	33.21
S1	41	28	1.49	0.51
S2	42	27	1.69	15.07
V1	39	30	123.18	29.50
V2	52	17	116.31	0.47

A.2   Primary Biliary Cirrhosis Data Set

The PBC data is from a clinical trial to evaluate the treatment for the primary biliary cirrhosis (PBC) of the liver. PBC is a rare and fatal chronic liver disease of unknown cause, with a prevalence of about 50-cases-per-million in the population. The study and the data set is described in the book, Fleming and Harrington (2005). The study involves 418 PBC patients who were recruited for the randomized study. Only 312 agreed to be randomized to placebo or to the drug D-penicillamine (treatment) and the remaining 106 agreed to be followed to obtain the primary outcome: the survival time and subjects were censored at the time of liver transplantation, lost-to-follow up or the end of the study. Numerous covariates were measured at the baseline but not all were available on every subject. Tables A.3 and A.4 provide the list of variables and their descriptive statistics, respectively. See Raghunathan, (2016) and appropriate sections of this book for examples of use.

Table A.3: Key Variables in Primary Biliary Cirrhosis Data Set

Variable Name	Label
Age	Age in Days
Albumin	Albumin in gm/dlG
Alk_phos	Alkaline phosphatase in U/liter
Ascites	Presence of ascites: 0=no 1=yes
Bili	Serum bilirubin in mg/dl
Chol	Serum cholesterol in mg/dl
Copper	Urine copper in ug/day
Drug	1= D-penicillamine, 2=placebo
Edema	Presence of edema 0=no edema or no diuretic therapy, .5 = edema present without diuretics, or edema resolved by diuretics, 1 = edema despite diuretic therapy
Futime	Number of days between registration and the earlier of death, transplantion, or study analysis time in July, 1986
Hepato	Presence of hepatomegaly 0=no 1=yes
ID	Case number
Platelet	Platelets per cubic ml/1000
Protime	Prothrombin time in seconds
Sex	0=male, 1=female
Sgot	SGOT in U/ml
Spiders	Presence of spiders 0=no 1=yes
Stage	Histologic stage of disease
Status	0=alive, 1=liver transplant, 2=dead
Trig	Triglicerides in mg/dl

Table A.4: Descriptive Statistics for the PBC Data

Variable	N	N Missing	Mean	Standard Deviation
ID	418	0	209.50	120.81
Futime	418	0	1917.78	1104.67
Status	418	0	0.83	0.96
Edrug	312	106	1.49	0.50
Age	418	0	18533.35	3815.85
Sex	418	0	0.89	0.31
Ascites	312	106	0.08	0.27
Hepato	312	106	0.51	0.50
Spiders	312	106	0.29	0.45
Edema	418	0	0.10	0.25
Bili	418	0	3.22	4.41
Chol	284	134	369.51	231.94
Albumin	418	0	3.50	0.42
Copper	310	108	97.65	85.61
Alk_phos	312	106	1982.66	2140.39
Sgot	312	106	122.56	56.70
Trig	282	136	124.70	65.15
Platelet	407	11	257.02	98.33
Protime	416	2	10.73	1.02
Stage	412	6	3.02	0.88

A.3 Opioid Detoxification Data Set

A randomized clinical trial was conducted to evaluate buprenorphine-naloxone (Bup-nx) versus clonidine for opioid detoxification (Ling et al, 2005). A total of 113 in-patients passing a screening test and satisfying inclusion criteria were randomized with 77 patients receiving Bup-nx and 36 receiving clonidine. The study involved measuring patients for 14 days with a base-line measure coded as day 0 (before the treatment began). A variety of measures were collected to assess the treatment effect with one of them a self-report measure, a visual analog scale (VAS) based on the question, "how much do you currently crave for opiates?" with continuous response options ranging from 0 (no cravings) to 100 (most extreme cravings possible). Not every patient responded to this question on all days. Tables A.5 and A.6 provide variable list, descriptions and summary statistics.

Table A.5: Key Variables of the Opioid Detoxification Data Set

Variable Name	Label
Age	Age in AGEU at RFSTDTC
Usubjid	Unique Subject Identifier
Visitnum	Visit Number
Day	Day of Visit
Female	Female
Instudy	In Study
Treat	1=Bup-Nx 0=Clonidine
Vas	Visual Analog Scale 0-10
White	White

Table A.6: Descriptive Statistics for the Opioid Detoxification Data Set

Variable	N	N Missing	Mean	Standard Deviation
Age	113	0	36.23	9.74
Treat	113	0	0.68	0.47
Female	113	0	0.40	0.49
White	113	0	0.56	0.50
Instudy	113	0	1.00	0
Num_visits	1582 person-visits	320 person-visits	10.65	4.84

A.4   American Changing Lives (ACL) Data Set

The Americans' Changing Lives (ACL) study is the oldest ongoing nationally representative longitudinal study of the role of a broad range of social, psychological, and behavioral factors (along with aspects of medical care and environmental exposure) in health and the way health changes with age over the adult life course. The study began in 1986 with a national face-to-face survey of 3,617 adults ages 25 and up in the continental U.S., with African Americans and people aged 60 and over over-sampled at twice the rate of the others, and face-to-face re-interviews in 1989 of 83% (n=2,867) of those still alive. Survivors have been re-interviewed by telephone, and where necessary face-to-face, in 1994 and 2001/02, and again in 2011/12. (http://www.isr.umich.edu/acl). Tables A.7 and A.8 provide variable list, descriptions and summary statistics.

A.5   National Comorbidity Survey Replication (NCS-R)

The National Comorbidity Survey Replication (NCS-R) is a probability sample of the United States carried out a decade after the original 1992 NCS (NCS-1) was conducted. The NCS-R repeats many of the questions from the NCS-1 and also expands the questioning to include assessments based on the diagnostic criteria of the American Psychiatric Association as reported in the Diagnostic and Statistical Manual - IV (DSM-IV), 1994. The survey was administered in two parts. Part I included a core diagnostic assessment of all respondents (n=9,282) that took an average of about 1 hour to administer. Part II included questions about risk factors, consequences, other correlates, and additional disorders. In an effort to reduce respondent burden and control study costs, Part II was administered only to 5,692 of the 9,282 Part I respondents, including all Part I respondents with a lifetime disorder plus a probability sub-sample of other respondents. The two major aims of the NCS-R were first, to investigate time trends and their correlates over the decade of the 1990s, and second, to expand the assessment in the baseline NCS-1 in order to address a number of important substantive and methodological issues that were raised by the NCS-1. See www.hcp.med.harvard/ncs for details. Tables A.9 and A.10 provide variable list, descriptions and summary statistics.

Table A.7: Key Variables of the ACL Data Set

Variable Name	Label
V1801	X1:SEX OF RESPONDENT
V2000	RA0C1(1):RESPONDENT AGE
V2064	4-CAT SOCIO-ECON STATUS
V2102	W1.Respondent Race/Ethnicity, 5-Category
Bh	Black and High Socio-Economic Status at 1986
Bl	Black and Low Socio-Economic Status at 1986
Blm	Black and Low-Middle Socio-Economic Status at 1986
Bum	Black and Upper Middle Socio-Economic Status at 1986
Caseid	CaseID
I1	Impairment Status at Wave 1 1986: 1=Moderate-Severe Impairment and Death 0=No Impairment
I2	Impairment Status at Wave 2 1989: 1=Moderate-Severe Impairment and Death 0=No Impairment
I3	Impairment Status at Wave 3 1994: 1=Moderate-Severe Impairment and Death 0=No Impairment
I4	Impairment Status at Wave 4 2001/2002: 1=Moderate-Severe Impairment and Death 0=No Impairment
I5	Impairment Status at Wave 5 2011/2012: 1=Moderate-Severe Impairment and Death 0=No Impairment
Ewh	White and High Socio-Economic Status at 1986
Wl	White and Low Socio-Economic Status at 1986
Wlm	White and Low-Middle Socio-Economic Status at 1986
Wum	White and Upper Middle Socio-Economic Status at 1986

Table A.8: Descriptive Statistics for the ACL Data

Variable	N	N Missing	Mean	Standard Deviation
V1801	3361	0	1.63	0.48
V2000	3361	0	54.19	17.60
V2064	3361	0	2.18	0.99
V2102	3361	0	1.34	0.48
Bh	3361	0	0.02	0.14
Bl	3361	0	0.16	0.37
Blm	3361	0	0.11	0.31
Bum	3361	0	0.06	0.23
Caseid	3361	0	1781.17	1044.54
I1	3361	0	0.22	0.41
I2	2844	517	0.17	0.37
I3	2756	605	0.36	0.45
I4	2718	643	0.49	0.50
I5	2344	1017	0.55	0.50
Ewh	3361	0	0.09	0.29
Wl	3361	0	0.15	0.36
Wlm	3361	0	0.20	0.40
Wum	3361	0	0.22	0.41

A.6   National Health and Nutrition Examination Survey, 2011-2012 (NHANES 2011-2012)

The National Center for Health Statistics (NCHS), Division of Health and Nutrition Examination Surveys (DHNES), part of the Centers for Disease Control and Prevention (CDC), has conducted a series of health and nutrition surveys since the early 1960's. The National Health and Nutrition Examination Surveys (NHANES) were conducted on a periodic basis from 1971 to 1994. In 1999, NHANES became continuous. Every year, approximately 5,000 individuals of all ages are interviewed in their homes and complete the health examination component of the survey. The health examination is conducted in a mobile examination center (MEC); the MEC provides an ideal setting for the collection of high quality data in a standardized environment. (http://www.cdc.gov/nchs/nhanes). Tables A.11 and A.12 provide variable list, descriptions and summary statistics.

Table A.9: Key Variables of the NCS-R Data Set

Variable Name	Label
AGE	Age at Interview
CASEID	CASE IDENTIFICATION NUMBER
DSM_SO	DSM-IV Social Phobia (Lifetime)
ED4CAT	Education 1=0-11 2=12 3=13-15 4=16+ Yrs
MAR3CAT	Marital Status 1=Married 2=Previously Married 3=Never Married
NCSRWTLG	NCSR sample part 2 weight
NCSRWTSH	NCSR sample part 1 weight
OBESE6CA	1=<18.5 2=18.5-24.9 3=25-29.9 4=30-34.9 5=35-39.9 6=40+
REGION	1=Northeast 2=Midwest 3=South 4=West
SECLUSTR	SAMPLING ERROR CLUSTER
SESTRAT	SAMPLING ERROR STRATUM
SEX	Sex 1=Male 2=Female
WKSTAT3C	Work Status 3 categories
Ag4cat	Age 1=17-29 2=30-44 3=45-59 4=60+
Ald	Alcohol Dependence 1=Yes 0=No
Mde	Major Depressive Episode 1=Yes 0=No
Racecat	Race 1=Other/Asian 2=Hispanic/Mexican 3=Black 4=White
Sexf	Female 1=Yes 0=No
Sexm	Male 1=Yes 0=No

Table A.10: Descriptive Statistics for the NCS-R Data

Variable	N	N Missing	Mean	Standard Deviation
CASEID	9282	0	4641.50	2679.63
DSM_SO	9282	0	4.51	1.31
AGE	9282	0	44.73	17.50
REGION	9282	0	2.57	1.01
MAR3CAT	9282	0	1.64	0.81
ED4CAT	9282	0	2.66	1.02
OBESE6CA	9106	176	2.93	1.11
NCSRWTSH	9282	0	1.00	0.52
NCSRWTLG	5692	3590	1.00	0.96
SEX	9282	0	1.55	0.50
WKSTAT3C	6633	2649	1.59	0.87
SESTRAT	9282	0	26.31	11.31
SECLUSTR	9282	0	1.51	0.50
Ag4cat	9282	0	2.44	1.06
Racecat	9282	0	3.52	0.86
Mde	9282	0	0.20	0.40
Ald	9282	0	0.05	0.21
Sexf	9282	0	0.55	0.50
Sexm	9282	0	0.45	0.50

Table A.11: Key Variables of the NHANES Data Set

Variable Name	Label
BMXBMI	Body Mass Index (kg/m**2)
BPXDI1	Diastolic: Blood pres (1st rdg) mm Hg
BPXDI2	Diastolic: Blood pres (2nd rdg) mm Hg
BPXDI3	Diastolic: Blood pres (3rd rdg) mm Hg
BPXDI4	Diastolic: Blood pres (4th rdg) mm Hg
BPXSY1	Systolic: Blood pres (1st rdg) mm Hg
BPXSY2	Systolic: Blood pres (2nd rdg) mm Hg
BPXSY3	Systolic: Blood pres (3rd rdg) mm Hg
BPXSY4	Systolic: Blood pres (4th rdg) mm Hg
DMDMARTL	Marital status
LBXTC	Total Cholesterol(mg/dL)
RIAGENDR	Gender
RIDRETH1	1=mex 2=oth hisp 3=white 4=black 5=other
RIDSTATR	Interview/Examination status
SDMVPSU	Masked variance pseudo-PSU
SDMVSTRA	Masked variance pseudo-stratum
SEQN	Respondent sequence number
WTINT2YR	Full sample 2 year interview weight
WTMEC2YR	Full sample 2 year MEC exam weight
Age	Age at Interview in Years
Age18p	Age 18+ 1=Yes 0=No
Black	Black
Bp_cat	Blood Pressure 1=Normal 2=Pre-Hypertension 3=Hypertension Stage 1 4=Hypertension Stage 2
Edcat	1=0-11 2=12 3=13-15 4=16+ Years of Education
Marcat	1=married 2=prev married 3=never married
Mex	Mexican
Other	Other Race/Ethnicity

Table A.12: Descriptive Statistics for the NHANES Data

Variable	N	N Missing	Mean	Standard Deviation
SEQN	9756	0	67038.50	2816.46
RIDSTATR	9756	0	1.96	0.20
RIAGENDR	9756	0	1.50	0.50
RIDRETH1	9756	0	3.23	1.25
DMDMARTL	5560	4196	2.75	3.34
WTINT2YR	9756	0	31425.86	34062.12
WTMEC2YR	9756	0	31425.86	35200.45
SDMVPSU	9756	0	1.64	0.64
SDMVSTRA	9756	0	95.87	3.98
BPXSY1	6756	3000	119.17	18.75
BPXDI1	6756	3000	66.90	15.11
BPXSY2	6908	2848	118.70	18.58
BPXDI2	6908	2848	66.28	16.07
BPXSY3	6917	2839	118.20	18.30
BPXDI3	6917	2839	65.91	16.63
BPXSY4	447	9309	118.99	21.47
BPXDI4	447	9309	71.78	14.93
BMXBMI	8602	1154	25.34	7.72
LBXTC	6988	2768	183.20	41.42
Age18p	9756	0	0.60	0.49
Edcat	8154	1602	2.10	1.16
Age	9756	0	31.40	24.58
Marcat	5553	4203	1.65	0.81
Bp_cat	7055	2701	1.55	0.71
Mex	9756	0	0.14	0.35
Black	9756	0	0.28	0.45
Other	9756	0	0.17	0.38

A.7 Health and Retirement Study, 2012 (HRS 2012)

The University of Michigan Health and Retirement Study (HRS) is a longitudinal panel study that surveys a representative sample of approximately 20,000 Americans over the age of 50 every two years. See hrsonline.isr.umich.edu for more details. Supported by the National Institute on Aging (NIA U01AG009740) and the Social Security Administration, the HRS explores the changes in labor force participation and the health transitions that individuals undergo toward the end of their work lives and in the years that follow. Since its launch in 1992, the study has collected information about income, work, assets, pension plans, health insurance, disability, physical health and functioning, cognitive functioning, and health care expenditures. Through its unique and in-depth interviews, the HRS provides an invaluable and growing body of multidisciplinary data that researchers can use to address important questions about the challenges and opportunities of aging. Tables A.13 and A.14 provide variable list, descriptions and summary statistics.

Table A.13: Key Variables of the HRS Data Set

Variable Name	Label
GENDER	Gender 1=Male 2=Female
H11ATOTA	H11ATOTA:W11 Total of all Assets–Cross-wave
H11ITOT	H11ITOT:W11 Incm: Total HHold / R+Sp only
HHID	HOUSEHOLD IDENTIFICATION NUMBER
NFINR	2012 WHETHER FINANCIAL RESPONDENT
NWGTHH	2012 WEIGHT: HOUSEHOLD LEVEL
NWGTR	2012 WEIGHT: RESPONDENT LEVEL
PN	RESPONDENT PERSON IDENTIFICATION NUMBER
SECU	SAMPLING ERROR COMPUTATION UNIT
STRATUM	STRATUM ID
Age65p	1=Age 65+ 0=Under Age 65
Arthritis	Arthritis 1=Yes 0=No
Diabetes	1=Yes Diabetes 0=No Diabetes
Edcat	Education 1=0-11 Yrs 2=12 Yrs 3=13-15 Yrs 4=16+ Yrs
Marcat	Marital Status 1=Married 2=Previously Married 3=Never Married
Numfalls24	Number of Falls Past 2 Years
Racecat	Race 1=Hispanic 2=NH White 3=NH Black 4=NH Other

Table A.14: Descriptive Statistics for the HRS Data

Variable	N	N Missing	Mean	Standard Deviation
NFINR	20554	0	2.24	1.85
GENDER	20553	1	1.58	0.49
SECU	20554	0	1.49	0.50
STRATUM	20554	0	29.89	15.70
NWGTHH	20554	0	4338.55	3654.94
NWGTR	20554	0	4412.71	3923.57
H11ATOTA	20554	0	375979.07	991439.87
H11ITOT	20554	0	61454.92	99654.17
Marcat	20542	12	1.47	0.59
Edcat	20454	100	2.49	1.07
Racecat	20517	37	2.12	0.67
Diabetes	20536	18	0.24	0.43
Numfalls24	10595	9959	1.19	3.32
Age65p	20554	0	0.52	0.50
Arthritis	20527	27	0.57	0.50

A.8 Case Control Data for Omega-3 Fatty Acids and Primary Cardiac Arrest

A case-control study was conducted to assess the relationship between dietary intake of omega-3 fatty acids and primary cardiac arrest (PCA) (defined as a sudden pulseless condition in the absence of any prior history of heart disease). In particular, the two fatty acids docosahexaenoic acid (DHA) and eicosapentaenoic (EPA) are of interest as they are not synthesized by the body and mostly derived through dietary intake of fish.

Briefly, a population-based case-control study was conducted in Seattle and King County, WA. All case subjects with primary cardiac arrest, aged 25 to 74 years, attended by paramedics during 1988 to 1994 (n = 334) were identified. Control subjects were randomly identified from the same defined population, matched by age (within 7 years) and sex (n = 493). Case and control subjects with prior clinically recognized heart disease or other major life-threatening morbidity and those who had taken fish-oil supplements during the prior year were excluded. All subjects were married and were residents of King County; their spouses participated in in-home interviews. (Siscovick et al, 1995, 2000). Tables A.15 and A.16 provide variable list, descriptions and summary statistics.

Table A.15: Key Variables of the PCA and Omega-3 Fatty Acids Data

Variable	Label
Age	Age in Years
Casecnt	Primary Cardiac Arrest Case Status 1=Case 0=Control
DHA_EPA	DHA EPA from Seafood Intake Study
Gender	Gender 0=Male, 1=Female
Redtot	Red Blood Cell Membrane EPA and DHA
StudyID	Study ID
Alcohol3	Alcohol Intake, Drinks Per Day
Cafftot	Caffeine Intake, MG Per Day
Cholesth	Hyper-Cholesterol
Diab	Diabetes
Edusubj3	Education 1=HS+,0=<HS
FAMMI	Family History of MI
FATINDEX	Fat Index Score
HGTCM	Height
HYPER	Hypertension
NUMCIG	Number of Cigarettes Per Day
RACE3	Race 1=White, 0=Non-White
SMOKE	Smoke Status 1=Not Smoker, 2=Former Smoker, 3=Current Smoker
TOTLKCAL	Kilo-calories Per Day
WGTKG	Weight
YRSSMOKE	Years Smoked

Table A.16: Descriptive Statistics for the PCA/Omega-3 Data

Variable	N	N missing	Mean	Standard Deviation
Casecnt	898	0	0.39	0.49
Age	898	0	58.63	10.19
Gender	898	0	0.21	0.41
Race3	897	1	0.94	0.24
HYPER	890	8	0.20	0.40
DIAB	894	4	0.07	0.26
SMOKE	896	2	1.80	0.75
NUMCIG	454	444	22.70	14.34
YRSSMOKE	489	409	28.16	14.67
FATINDEX	864	34	21.560	3.94
FAMMI	891	7	0.45	0.50
EDUSUBJ3	898	0	0.71	0.45
DHA_EPA	898	0	4.91	5.73
REDTOT	498	400	4.65	1.17
CHOLESTH	884	14	0.23	0.42
CAFFTOT	896	2	381.701	462.21
WGTKG	845	53	81.32	16.36
TOTLKCAL	898	0	1195.84	1587.40
ALCOHOL3	897	1	0.92	1.74
HGTCM	896	2	176.22	9.23

A.9   National Merit Twin Study

The National Merit Twin Study is a data set focused on the study of twins and test scores from the National Merit tests. The data used in this book is at the individual level, thus each case represents an unique person. For details and general documentation, see Loehlin,J.C. & Nichols, R.C. (1976). Genes, Environment and Personality. Austin TX: University of Texas Press. Tables A.17 and A.18 provide variable list, descriptions and summary statistics.

Table A.17: Key Variables of the National Merit Twin Study Data

Variable Name	Label
Pairnum	Twin pair number
Sex	1=Male 2=Female
Zygosity	Twin Status 1=Identical 2=Fraternal
Moed	Mother's educational level (coded 1-6, see codebook)
Faed	Father's educational level (coded 1-6, see codebook)
Finc	Family income level (coded 1-7, see codebook)
English	NMSQT Subtest: English
Math	NMSQT Subtest: Mathematics
SocSci	NMSQT Subtest: Social Science
NatSci	NMSQT Subtest: Natural Science
Vocab	NMSQT Subtest: Vocabulary

Table A.18: Descriptive Statistics for the National Merit Twin Study Data

Variable	N	NMiss	Mean	Standard Deviation
Pairnum	1678	0	761.23	10.63
Sex	1678	0	1.582	0.01
Zygosity	1678	0	1.39	0.01
Moed (Mothers Education)	1640	38	3.42	0.03
Faed (Fathers Education)	1630	48	3.60	0.04
Finc (Family Income)	1554	124	3.22	0.04
NMT: English	1678	0	19.68	0.11
NMT: Mathematics	1678	0	21.15	0.15
NMT: Social Science	1678	0	20.57	0.12
NMT: Natural Science	1678	0	19.99	0.14
NMT: Vocabulary	1678	0	20.96	0.12

A.10   European Social Survey-Russian Federation

The European Social Survey (ESS) is an academically driven cross-national survey that has been conducted every two years across Europe since 2001. The survey measures the attitudes, beliefs and behaviour patterns of diverse populations in more than thirty nations. The main aims of the ESS are to chart stability and change in social structure, conditions and attitudes in Europe and to interpret how Europe's social, political and moral fabric is changing to achieve and spread higher standards of rigour in cross-national research in the social sciences, including for example, questionnaire design and pre-testing, sampling, data collection, reduction of bias and the reliability of questions to introduce soundly-based indicators of national progress, based on citizens' perceptions and judgements of key aspects of their societies to undertake and facilitate the training of European social researchers in comparative quantitative measurement and analysis to improve the visibility and outreach of data on social change among academics, policy makers and the wider public. The ESS data is available free of charge for non-commercial use and can be downloaded from this website after a short registration. See http://www.europeansocialsurvey.org. The Russian Federation Round 6 data is based upon a complex sample and therefore, weights, stratification and PSU variables are included. Key questions about satisfied with present state of economy in country, satisfaction with life, and trust in the police plus demographic indicators are used for selected analyses. Tables A.19 and A.20 provide variable list, descriptions and summary statistics.

A.11   Outline of Analysis Examples and Data Sets

Table A.21 provides an outline of analysis examples and data sets used in this book.

Table A.19: Key Variables of the ESS Russian Federation Data

Variable Name	Label
AGEA	Age of Respondent
DWEIGHT	Design Weight
EISCED	Education level Europe
STFECO	Satisfaction with Present State of Economy
STFLIFE	Satisfaction with Life
TRSTPLC	Trust in Police
TVTOT	Hours of TV Avg. Wkday: 0=None 1 le 0.5 hrs 2=.5-1 hrs 3=1-1.5 hrs 4=1.6-2 hrs 5=2.1-2.5 hrs 6=2.6-3 hrs 7 ge 3 hrs
AGECAT	Age in Categories
IDNO	Respondent ID
MALE	Male
MARCAT	Marital Status
PSU	Primary Sampling Unit
STRATIFY	Stratification

Table A.20: Descriptive Statistics for the ESS Russian Federation Data

Variable	N	NMiss	Mean	Standard Deviation
IDNO	2484	0	4354.48	2414.53
PSU	2484	0	71013.57	163694.84
TVTOT	2426	58	4.39	2.17
TRSTPLC	2390	94	3.49	2.56
VOTE	2464	20	1.39	0.57
STFLIFE	2458	26	5.79	2.33
STFECO	2362	122	3.81	2.25
GNDR	2484	0	1.62	0.49
AGEA	2478	6	45.94	18.05
EISCED	2484	0	4.97	1.67
DWEIGHT	2484	0	1.00	0.58
MALE	2484	0	0.38	0.49
AGECAT	2478	6	2.54	1.11
MARCAT	2444	40	1.80	0.80

Table A.21: Data Sets Used in Various Examples

Analytic Technique	Data Set
Descriptive Analyses of Continuous and Categorical Variables	NHANES 2011-2012
Linear Regression	NHANES 2011-2012
Logistic Regression (Binary and Multinomial)	Primary Cardiac Arrest, NCS-R
Count Models	HRS 2012
Structural Equation Models	National Merit Twin Data
Categorical Models	Data from Little and Rubin (2002, Table 13.1 and Table 13.8) and Stokes, Davis, and Koch (2000)
Longitudinal Data Analysis	Opioid Detoxification Data, Panel Study of Income Dynamics (see Chapter 8, Longitudinal Data Analysis for details on download and construction of this data set), and American Changing Lives
Survival Analysis	Primary Biliary Cirrhosis Data, and Academic Aptitude Data
Complex Survey Data Analysis using BBDesign	NHANES 2011-2012
Sensitivity Analysis	Opioid Detoxification Data, Primary Cardiac Arrest, ESS6 Russian Federation Data

B

IVEware

IVEware

B.1   What is *IVEware*?

IVEware (Version 0.3) is a collection of routines written under various platforms and packaged to perform multiple imputations and analysis of multiply imputed data sets. The software can also be used to perform analysis without missing data. *IVEware* defaults to assuming a simple random sample, but uses the Jackknife Repeated Replication or Taylor Series Linearization techniques for analyzing data from complex surveys.

IVEware can be run with SAS, Stata, R, SPSS or as stand-alone under the Windows or Linux environment. The R, Stata, SPSS and the stand-alone versions can also be used with the Mac OS. The stand-alone version has limited capabilities for analyzing the multiply imputed data sets though the routines for creating imputations are the same across all packages. The command structure is the same across all platforms. The most preferable way to execute *IVEware* is to use the built-in XML editor. However, it can also be run using the built-in editor within the four software packages previously mentioned. Throughout the book, the built-in XML editor is used in all the examples and the reader can refer to the user manual (which can be downloaded from the web site www.iveware.org) for examples using the built-in editors within the four software packages.

The user can also mix and match the codes from these software packages through a standard XML toggle-parser. For example, the following code

```
<sas name="myfile">
    SAS Commands here
</sas>
<R name="myfile">
    R-commands here
</R>
```

will execute the *SAS commands* and store the commands in the file "*myfile.sas*" and execute the R-commands and store them in the file "*myfile.R*".) if the provided XML editor is used to execute *IVEware* commands.

B.2   Download and Setup

IVEware can be downloaded and installed from **www.iveware.org**. Installation instructions and setup are slightly different for Windows, Linux and MAC operating systems, therefore, it is very important to follow the installation instructions for *IVEware* to work properly.

B.2.1   Windows

Download the file **srclib_windows.exe** and double click on it to install. By default, all the relevant software files are extracted to the directory

```
C:\Program Files (x86)\srclib
```

You may change the directory and the location you choose will then replace

```
~/srclib
```

in the guides given later to make *IVEware* work with one or more packages SAS, SPSS, Stata, and R. The installer will automatically create a desktop icon unless you choose not to.

The next step is to make sure that *IVEware* can execute the appropriate software. Find the path of the directory where the executable file of the software is located. For example, the typical location of the file "sas.exe" to run SAS version 9.4 is,

```
C:\Program Files\SASHome2\SASFoundation\9.4\sas.exe
```

Similarly, a 64-bit executable "Rgui.exe' may be located in

```
C:\Program Files\R\R-3.2.3\bin\x64\Rgui.exe
```

Find the path of the other software executable files, if needed.

Next, edit the file "settings.xml" in the "srclib" directory to match your own paths. Here is an example of a file appropriate to run *IVEware* with all four software (SAS, R, Stata, and SPSS) packages:

```
<settings>
<frameworks>
<sas path="C:\Program Files\SASHome2\SASFoundation\9.4\sas.exe" />
<spss path="C:\Program Files\IBM\SPSS\Statistics\22\stats.exe" />
<stata path="M:\stata14SE-64bit\StataSE-64.exe" />
<R path="C:\Program Files\R\R-3.2.3\bin\x65\Rgui.exe" />
<gnuplot path="C:\Program Files\gnuplot\bin\wgnuplot.exe" />
</frameworks>
</settings>
```

In this example, the Stata executable file is located in the network drive M:. The stand-alone version uses "GNUPLOT" for creating and displaying graphics. You should download and install this package. For more details about this software, see **www.gnuplot.info**.

To verify that *IVEware* is correctly installed, download the example file **ive_examples_windows.zip** and extract it to the directory of your choice (for example, create a subdirectory called **iveware** under the **Documents** directory). Double click the *IVEware* icon which will open an XML editor window. Choose "File", navigate to the directory with the example files, choose any example XML file to open, and click "Run". You can also run *IVEware* in batch mode by typing:

```
"C:\Program Files (x86)\srclib\srcexec" ive_example_file.xml
```

in the command window after setting the current directory to the directory where the example file is located. Use MS Word or another software to check the "*.log" files produced by the run to see that there were no errors and also compare the "*.lst" files produced by run with the corresponding "*.chk" files. They should differ only in the dates.

B.2.2   Linux

For installing *IVEware* on a Linux system, download the file "srclib_pclinux64.tgz" from www.iveware.org and extract the srclib directory into an appropriate parent directory, such as, "/usr/local/" or in your home directory. You can copy the files into the subdirectory of your choice. The location you choose will replace '~ /*srclib*" in guides for using *IVEware* with R, SAS, SPSS, Stata and SRCware, as described in later examples. If you plan to use Srclib with R, SAS, SPSS or Stata and the version cannot be invoked by its lower-case name, edit the "srclib/settings.xml" file to provide the correct path.

As before, you can verify that *IVEware* is installed correctly by downloading the file "ive_examples_pclinux.tgz" and extracting the examples directory into an appropriate parent directory, for example, the home directory. Navigate to the srclib directory, double-click the srcshell icon, click File-Open, navigate to the Examples directory, open an appropriate setup file, for example, "ive_examples_srcware.xml", and finally, click Run. You can also run the program in batch mode. For example, navigate to the examples directory and use srcexec to run an appropriate setup file:

```
~/srclib/bin/srcexec ive_examples_sas.xml
```

where ~ /*srclib* is the directory containing *IVEware*.

Use the Linux "cat" and "diff" commands (or other software such as OpenOffice), check the *.log files produced by the run to see that there were no errors and compare the *.lst files produced by run with the corresponding *.chk files. Again, they should differ only in the dates.

B.2.3   Mac OS

For those using a MAC-OS, *IVEware* can be used only as stand-alone or with Stata, R and SPSS. To install on a Mac, download the file "srcware_macosx64.tgz" from www.iveware.org and extract "Srcware.app" into an appropriate parent directory, say, */Applications/*, or the home directory. You can choose any location. A subdirectory of the location you choose will replace ∼ */srclib* in the guides for using *IVEware* with R, SPSS, Stata and Srcware.

As in the case of Linux, if you plan to use Srclib with R, SPSS or Stata, note that the syntax is case sensitive and edit the file

`~/Srcware.app/Contents/MacOS/settings.xml`,

to provide the correct path.

Once again, you can verify that the Srcware app is installed correctly. Download the file "ive_examples_macosx.tgz" from www.iveware.org, extract it to any directory (Examples) of your choice. Double-click Srcware.app to run the app, click File - Open, navigate to the Examples directory, open an appropriate setup file, for example, "ive_examples_srcware.xml", and click Run. *IVEware* can be run in batch mode by navigating to the examples directory and submit srcexec to run an appropriate setup file, for example,

`~/Srcware.app/Contents/MacOS/srcexec ive_examples_sas.xml`

B.3   Structure of *IVEware*

IVEware is organized into seven modules to perform various tasks. The six core modules are **IMPUTE, BBDESIGN, DESCRIBE, REGRESS, SYNTHESIZE** and **COMBINE** and the seventh module, **SASMOD,** is specific to SAS.

1. **IMPUTE** uses a multivariate sequential regression approach (Raghunathan et al (2001), Raghunathan (2016)). This approach is also called Chained Equations, (Van Buuren and Oudshoorn (1999)) and Fully Conditional Specification (Van Buuren (2012)) and is used to impute item missing values or unit non-response. IMPUTE can create multiply imputed data sets and can handle continuous, categorical, count and semi-continuous variables.

2. **BBDESIGN** implements the weighted finite population Bayesian Bootstrap approach to generate synthetic populations from complex survey data. The primary goal is to incorporate weighting, clustering and stratification using a nonparametric approach for generating

the non-sampled portion of the population from the posterior predictive distribution, conditional on the observed data and the design information. For more details see Zhou, Elliott and Raghunathan (2016a, 2016b, 2016c)

3. **DESCRIBE** estimates population means, proportions, subgroup differences, contrasts and linear combinations of means and proportions. For data from complex surveys, a Taylor Series Linearization approach is used to obtain variance estimates appropriate for a user-specified complex sample design.

4. **REGRESS** fits linear, logistic, polytomous, Poisson, Tobit and proportional hazard regression models. For data resulting from a complex sample design, the Jackknife Repeated Replication technique is used to obtain variance estimates.

5. **SASMOD** (requires SAS) allows users to take into account complex sample design features when analyzing data with selected SAS procedures. Currently the following SAS PROCS can be called: CALIS, CATMOD, GENMOD, LIFEREG, MIXED, NLIN, PHREG, and PROBIT.

6. **SYNTHESIZE** uses the multivariate sequential regression approach to create full or partial synthetic data sets to limit statistical disclosure (See Raghunathan, Reiter and Rubin (2003), Reiter (2002) and Little, Liu and Raghunathan (2004) for more details.) All item missing values are also imputed when creating synthetic data sets. However, DESCRIBE, REGRESS and SASMOD modules cannot be used to analyze synthetic data sets as they DO NOT implement the appropriate combining rules. Examples of implementation of correct combining rules for synthesized data sets are provided in the user manual available at the web site www.iveware.org.

7. **COMBINE** is useful for combining information from multiple sources through multiple imputation. Suppose that Data 1 provides variables X and Y, Data 2 provides variables X and Z and Data 3 provides variables Y and Z. COMBINE can be used to concatenate the three data sets and multiply impute the missing values of X, Y and Z to create large data sets with complete data on all three variables. All item missing values in the individual data sets will also be imputed. The multiply imputed combined data sets can be analyzed using DESCRIBE, REGRESS and SASMOD modules (see Schenker, Raghunathan, and Bondarenko (2010) for an application and Dong, Elliott and Raghunathan (2014a) for more details).

There are many packages such as R ("with" and "pool"), Stata ("mi estimate"), and SAS ("PROC MI, PROC MIANALYZE") to analyze multiply imputed data sets. All these packages can be used within the "XML" structure of *IVEware*.

A more detailed manual of operation on how to use *IVEware* can be downloaded from the web site **www.iveware.org**. The web site also contains numerous tutorials with examples, codes and output under all versions of *IVEware*. For more details about the technical aspects, see Raghunathan (2016).

Bibliography

[1] Abayomi, K., Gelman, A., & Levy, M. (2008). Diagnostics for multivariate imputations. *Journal of the Royal Statistical Society: Series C (Applied Statistics)*, 57, 273-291.

[2] Afifi, A. A., & Elashoff, R. M. (1967). Missing observations in multivariate statistics: II. point estimation in simple linear regression. *Journal of the American Statistical Association, 62*, 10-29.

[3] Agresti, A. (2012). *Categorical data analysis* (3rd ed.) Wiley.

[4] Allan, F. E., & Wishart, J. (1930). A method of estimating the yield of a missing plot in field experimental work. *Jour. Agr. Sci., 20*, 399-406.

[5] Allison, P. (2012). Why you probably need more imputations than you think." accessed february 20, 2015, http://www.statisticalhorizons.com/more-imputations.

[6] Allison, P. D. (2002). *Quantitative applications in the social sciences: Missing data*. Thousand Oaks, CA: SAGE Publications Ltd.

[7] Allison, P. D. (2003). Missing data techniques for structural equation modeling. *Journal of Abnormal Psychology, 112*, 545-557.

[8] Amemiya, T. (1984). Tobit models: A survey. *Journal of Econometrics, 24*(1-2), 3-61.

[9] Baker, S. G., & Laird, N. M. (1988). Regression analysis for categorical variables with outcome subject to nonignorable nonresponse. *Journal of the American Statistical Association, 83*, 62-69.

[10] Barnard, J., & Rubin, D. B. (1999). Miscellanea. small-sample degrees of freedom with multiple imputation. *Biometrika, 86*, 948-955.

[11] Bartlett J.W., Seaman S.R., White I.R., Carpenter J.R., & Alzheimer's Disease Neuroimaging Initiative (2015). Multiple imputation of covariates by fully conditional specification: Accommodating the substantive model. *Statistical Methods in Medical Research*, 24, 462-87.

[12] Berglund, P., & Heeringa, S. G. (2014). *Multiple imputation of missing data using SAS* SAS Institute.

[13] Bickel, P. J.,Doksum, K. A. (2006). *Mathematical statistics: Basic ideas and selected topics* (2nd ed ed.). Upper Saddle River, NJ: Pearson Prentice Hall.

[14] Bishop, Y. M., Fienberg, S. E., & Holland, P. W. (1975). *Discrete multivariate analysis: Theory and practice.* Cambridge, Mass: MIT Press.

[15] Bodner, T. E. (2008). What improves with increased missing data imputations? *Structural Equation Modeling, 15,* 651-675.

[16] Bollen, K. A. (1989). *Structural equations with latent variables.* Wiley.

[17] Bondarenko, I., & Raghunathan, T. E. (2010). Multiple imputation for causal inference. *Arbor, 1001*(48109), 48109.

[18] Bondarenko, I., & Raghunathan, T. E. (2016). Graphical and numerical diagnostic tools to assess suitability of multiple imputations and imputation models. *Statistics in Medicine, 35,* 3007-3020.

[19] Box, G. E. P., & Tiao, G. C. (1973). *Bayesian inference in statistical analysis.* New York: Wiley Classics.

[20] Brand, J. P. L. (1999). *Development, implementation and evaluation of multiple imputation strategies for the statistical analysis of incomplete data sets.* (Ph.D. Thesis, Erasmus University, Rotterdam).

[21] Carlin, B. P., & Louis, T. A. (2008). *Bayesian methods for data analysis* (3rd ed. ed.). Boca Raton, FL: CRC Press, Taylor & Francis Group.

[22] Carpenter, J., & Kenward M. (2013). *Multiple imputation and its application* Wiley.

[23] Casella, G., & Berger, R. L. (2002). In Crockett C. (Ed.), *Statistical inference* (2nd ed.). Pacific Grove, CA, United States: Duxbury Press.

[24] Chambers, R. L., & Skinner, C. J. (2003). *Analysis of survey data.* New York: Wiley.

[25] Cochran, W. G. (1977). *Sampling techniques* (3rd ed. ed.). New York: John Wiley & Sons.

[26] Cox, D. R., & Hinkley, D. V. (1979). *Theoretical statistics.* Boc a Raton, FL: CRC Press.

[27] Davidian, M., Tsiatis, A. A., & Leon, S. (2005). Semiparametric estimation of treatment effect in a Pretest–Posttest study with missing data. *Statistical Science, 20,* 24 August 2005-261-301.

[28] Dempster, A. P., Rubin, D. B., & Tsutakawa, R. K. (1981). Estimation in covariance components models. *Journal of the American Statistical Association, 76,* 341-353.

[29] Diggle, P., Heagerty, P., Liang, K., & Zeger, S. (2002). *Analysis of longitudinal data*. Oxford, UK: Oxford University Press.

[30] Dodge, Y. (1985). *Analysis of experiments with missing data*. New York: John Wiley & Sons.

[31] Dong, Q., Elliott, M. R., & Raghunathan, T. E. (2014a). Combining information from multiple complex surveys. *Survey Methodology, 40*, 347-354.

[32] Dong, Q., Elliott, M. R., & Raghunathan, T. E. (2014b). A nonparametric method to generate synthetic populations to adjust for complex sampling design features. *Survey Methodology*, 40(1), 29-46.

[33] Draper, N. R., & Smith, H. (1998). *Applied regression analysis* (3rd ed.). New York: Wiley.

[34] Enders, C. K. (2006). A primer on the use of modern missing-data methods in psychosomatic medicine research. *Psychosomatic Medicine, 68*, 427-436.

[35] Enders, C. K. (2010). In Kenny D. A., Little T. D. (Eds.), *Applied missing data analysis*. New York: The Guilford Press.

[36] Engels, J. M., & Diehr, P. (2003). Imputation of missing longitudinal data: A comparison of methods. *Journal of Clinical Epidemiology, 56*, 968-976.

[37] Fay, R. E. (1992). When are inferences from multiple imputation valid? *Proceedings of the Survey Research Methods Section of the American Statistical Association,* pp. 227-232.

[38] Fleming, T. R., & Harrington, D. P. (2005). *Counting processes and survival analysis*. Hoboken, New Jersey: John Wiley & Sons.

[39] Gadbury, G. L., Coffey, C. S., & Allison, D. B. (2003). Modern statistical methods for handling missing repeated measurements in obesity trial data: Beyond LOCF. *Obesity Reviews : An Official Journal of the International Association for the Study of Obesity, 4*(3), 175-184.

[40] Gelman, A., Carlin, J. B., Stern, H. S., & Rubin, D. B. (2004). *Bayesian data analysis*. Boca Raton, FL: Chapman and Hall/CRC.

[41] Gelman, A., & Hill, J. (2006). *Data analysis using regression and Multilevel/Hierarchical models*. New York: Cambridge University Press.

[42] Gelman, A., Carlin, J. B., Stern, H. S., Dunson, D. B., Vehtari, A., & Rubin, D. B. (2013). *Bayesian data analysis, third edition*. Boca Raton, FL: CRC Press, Taylor & Francis Group.

[43] Glynn, R., Laird, N., & Rubin, D. B. (1986). Selection modeling versus mixture modeling with nonignorable nonresponse. In H. Wainer (Ed.), (pp. 115-142) Springer New York.

[44] Graham, J. W., Olchowski, A. E., & Gilreath, T. D. (2007). How many imputations are really needed? some practical clarifications of multiple imputation theory. *Prevention Science : The Official Journal of the Society for Prevention Research, 8,* 206-213.

[45] Graubard, B. I., & Korn, E. L. (1999). Predictive margins with survey data. *Biometrics, 55,* 652-659.

[46] Grizzle, J. E., Starmer, C. F., & Koch, G. G. (1969). Analysis of categorical data by linear models. *Biometrics, 25,* 489-504.

[47] He, Y., & Raghunathan, T. E. (2006). Tukey's gh distibution for multiple imputation. *The American Statistician, 60,* 251-256.

[48] He, Y., & Raghunathan, T. E. (2012). Multiple imputation using multivariate gh transformations.

[49] He, Y., Zaslavsky, A. M., Harrington, D. P., Catalano, P., & Landrum, M. B. (2010). Multiple imputation in a large-scale complex survey: A practical guide. *Statistical Methods in Medical Research, 19,* 653-670.

[50] Heckman, J. J. (1976). The common structure of statistical models of truncation, sample selection and limited dependent variables and a simple estimator for such models. *Annals of economic and social measurement* (Sanford V. Berg ed., pp. 475-4925) NBER.

[51] Heeringa, S. G., West, B. T., & Berglund, P. A. (2017). *Applied survey data analysis.* Boca Raton, FL: Chapman & Hall/CRC.

[52] Heyting, A., Tolboom, J. T., & Essers, J. G. (1992). Statistical handling of drop-outs in longitudinal clinical trials. *Statistics in Medicine, 11,* 2043-2061.

[53] Hoaglin, D. C., Mosteller, F., & Tukey, J. W. (1985). *Exploring data tables, trends, and shapes.* New York: Wiley.

[54] Hogg, R. V., KcKean, J., & Craig, A. T. (2012). *Introduction to mathematical statistics* (7th ed.). U.S.A.: Pearson Education Limited.

[55] Hosmer, D., Lemeshow, S., & May, S. (2008). *Applied survival analysis: Regression modeling of time to event data* (2nd ed.) Wiley.

[56] Hosmer, D. W., Lemeshow, S., & Sturdivant, R. X. (2013). *Applied logistic regression* (3rd ed.). New Jersey: Wiley.

[57] House, J.,S. (2014). Americans' changing lives: Waves I, II, III, IV, and V, 1986, 1989, 1994, 2002, and 2011. *Ann Arbor, MI: Inter-University Consortium for Political and Social Research [Distributor], ICPSR04690-v7*

[58] Hughes, R. A., White, I. R., Seaman, S. R., Carpenter, J. R., Tilling, K., & Sterne, J. A. C. (2014) Joint modelling rationale for chained equations. *BMC Medical Research Methodology*, 14-28.

[59] Jonathan, W. B., Shaun, R. S., Ian, R. W., James, R. C., & for the Alzheimer's Disease,Neuroimaging Initiative. (2015). Multiple imputation of covariates by fully conditional specification: Accommodating the substantive model. *Stat Methods Med Res, 24*, 462-487.

[60] Kaciroti, N. A., & Raghunathan, T. E. (2014). Bayesian sensitivity analysis of incomplete data: Bridging pattern-mixture and selection models. *Statistics in Medicine, 33*, 4841-4857.

[61] Kalbfleisch, J. D., & Prentice, R. L. (2002). *The statistical analysis of failure time data* (2nd ed.). New York: John Wiley and Sons.

[62] Kennickell, A. B. (1991). Imputation of the 1989 survey of consumer finances: Stochastic relaxation and multiple imputation. *Proceedings of the Survey Research Methods Section of the American Statistical Association*, 1-10.

[63] Kim, J. K., Michael Brick, J., Fuller, W. A., & Kalton, G. (2006). On the bias of the multiple-imputation variance estimator in survey sampling. *Journal of the Royal Statistical Society: Series B (Statistical Methodology), 68*, 509-521.

[64] Kim, J. K., & Shao, J. (2014). *Statistical methods for handling incomplete data*. Boca Raton: CRC Press, Taylor & Francis Group.

[65] Kim, S., Sugar, C. A., & Belin, T. R. (2015). Evaluating model-based imputation methods for missing covariates in regression models with interactions. *Statistics in Medicine, 34*, 1876-1888.

[66] Kish, L., & Frankel, M. (1974). Inference from complex systems. *Journal of the Royal Statistical Society.Series B (Methodological), 36*, 1-37.

[67] Kish, L. (1965). *Survey sampling*. New York: John Wiley and Sons, Inc.

[68] Kline, R. B. (1998). *Principles and practice of structural equation modeling* Guilford Publications.

[69] Laird, N. M., & Ware, J. H. (1982). Random-effects models for longitudinal data. *Biometrics, 38*, 963-974.

[70] Lavori, P. W., Dawson, R., & Shera, D. (1995). A multiple imputation strategy for clinical trials with truncation of patient data. *Statistics in Medicine, 14*, 1913-1925.

[71] Li, K. H., Meng, X. L., Raghunathan, T. E., & Rubin, D. B. (1991b). Significance levels from repeated P-values with multiply-imputed data. *Statistica Sinica, 1*, 65-92.

[72] Li, K. H., Raghunathan, T. E., & Rubin, D. B. (1991a). Large-sample significance levels from multiply imputed data using moment-based statistics and an F-reference distribution. *Journal of the American Statistical Association, 86*, 1065-1073.

[73] Lillard, L., Smith, J. P., & Welch, F. (1986). What do we really know about wages? the importance of nonreporting and census imputation. *Journal of Political Economy, 94* (3, Part 1), 489-506.

[74] Ling, W., A., L., , S., Annon, J. J., Hillhouse, M., Babcock, D., et al. (2005). A multi-center randomized trial of buprenorphine-naloxone versus clonidine for opioid, detoxification: Findings from the national institute on drug abuse clinical trials network. *Addiction, 100*, 1090-1100.

[75] Little, R. J. A. (1982). Models for nonresponse in sample surveys. *Journal of the American Statistical Association, 77*, 237-250.

[76] Little, R. J. A. (1985). Nonresponse adjustments in longitudinal surveys: Models for categorical data. *Bulletin of the International Statistical Institute, Proceedings of the 45th Session: Invited Papers, , Section 15.1.* pp. 1-18.

[77] Little, R. J. A. (1992). Regression with missing X's: A review. *Journal of the American Statistical Association, 87*, 1227-1237.

[78] Little, R. J. A. (1993). Statistical analysis of masked data. *Journal of Official Statistics, 9*, 407-426.

[79] Little, R. J. A. (1994). A class of pattern-mixture models for normal incomplete data. *Biometrika, 81*, 471-483.

[80] Little, R. J. A. (1995). Modeling the drop-out mechanism in repeated-measures studies. *Journal of the American Statistical Association, 90*, 1112-1121.

[81] Little, R. J. A., Liu, F., & Raghunathan, T. E. (2004). Statistical disclosure techniques based on multiple imputation. *Applied bayesian modeling and causal inference from incomplete-data perspectives* (pp. 141-152) John Wiley & Sons, Ltd.

[82] Little, R. J. A., & Rubin, D. B. (1987). *Statistical analysis with missing data* (1st Edition ed.). New York: John Wiley & Sons.

[83] Little, R. J. A., & Rubin, D. B. (2002). *Statistical analysis with missing data* (2nd Edition ed.). New York: John Wiley & Sons.

[84] Little, R. J. A., & Schenker, N. (1995). Missing data. In G. Arminger, C. C. Clogg & M. E. Sobel (Eds.), *Handbook of statistical modeling for the social and behavioral sciences* (pp. 39-75) Springer Science+Business Media, LLC.

[85] Little, R. J. A., & Schluchter, M. D. (1985). Maximum likelihood estimation for mixed continuous and categorical data with missing values. *Biometrika, 72*, 497-512.

[86] Liu, C., & Rubin, D. B. (1998). Maximum likelihood estimation of factor analysis using the ecme algorithm with complete and incomplete data. *Statistica Sinica, 8*, 729-747.

[87] Liu, J., Gelman, A., Hill, J., Su, Y.S., & Kropko, J.. (2014). On the stationary distribution of iterative imputations. *Biometrika*, 101, 155-173.

[88] Loehlin, J. C., & Nichols, R. C. (1976). *Heredity, environment, and personality: A study of 850 sets of twins*. Austin: University of Texas Press.

[89] Lohr, S. L. (2009). *Sampling: Design and analysis* (2nd ed ed.). United States: Brooks/Cole, Cengage Learning.

[90] Long, J. S. (1997). *Regression models for categorical and limited dependent variables*. Thousand Oaks: Sage Publications.

[91] Long, J. S., & Freese, J. (2006). *Regression models for categorical dependent variables using stata, 2nd edition* StataCorp LP.

[92] Lumley, T. S. (2010). *Complex surveys: A guide to analysis using R*. New York: John Wiley & Sons.

[93] Mallinckrodt, C. H., Clark, W. S., & David, S. R. (2001). Accounting for dropout bias using mixed-effects models. *Journal of Biopharmaceutical Statistics, 11*(1-2), 9-21.

[94] Mallinckrodt, C. H., Sanger, T. M., Dube, S., DeBrota, D. J., Molenberghs, G., Carroll, R. J., et al. (2003). Assessing and interpreting treatment effects in longitudinal clinical trials with missing data. *Biological Psychiatry, 53*, 754-760.

[95] Marjoribanks, K. (1974). *Environments for learning* National Foundation for Educational Research Publications.

[96] McCullagh, P., & Neder, J. A. (1989). *Generalized linear models. (2nd Edition ed.)*. London: Chapman and Hall.

[97] Meng, X. (1994). Multiple-imputation inferences with uncongenial sources of input. *Statistical Science, 9*, 538-558.

[98] Meng, X., & Rubin, D. B. (1992). Performing likelihood ratio tests with multiply-imputed data sets. *Biometrika, 79*, 103-111.

[99] Molenberghs, G., & Verbeke, G. (2005). *Models for repeated discrete data*. U.S.A.: Springer.

[100] Molenberghs, G., & Kenward, M. G. (2007). *Missing data in clinical studies*. New York: Wiley & Sons.

[101] National Research Council. (2010). *The prevention and treatment of missing data in clinical trials. panel on handling missing data in clinical trials, committee on national statistics, division of behavioral and social sciences and education*. Washington, DC: The National Academies Press.

[102] Neter, J., Wasserman, W., Kutner, M., & Nachtsheim, C. (1996). *Applied linear statistical models* McGraw-Hill/Irwin.

[103] Nordheim, E. V. (1984). Inference from nonrandomly missing categorical data: An example from a genetic study on turner's syndrome. *Journal of the American Statistical Association, 79*, 772-780.

[104] O'Neill, ,R.T., & Temple, R. (2012). The prevention and treatment of missing data in clinical trials: An FDA perspective on the importance of dealing with it. *Clinical Pharmacology & Therapeutics, 91*, 550-554.

[105] Pregibon, D. (1977). Typical survey data: Estimation and imputation. *Survey Methodology, *, 70-102.

[106] Preisser, J. S., Lohman, K. K., & Rathouz, P. J. (2002). Performance of weighted estimating equations for longitudinal binary data with drop-outs missing at random. *Statistics in Medicine, 21*, 3035-3054.

[107] Raghunathan, T. E., Solenberger, P., Berglund, P., & VanHoewyk, J. (2017). IVEware: Imputation and variance estimation software.*(version 0.3)*, www.iveware.org.

[108] Raghunathan, T. E. (1987). Large sample significance levels from multiply-imputed data. Harvard University).

[109] Raghunathan, T. E. (2016). *Missing data analysis in practice*. Boca Raton: CRC Press.

[110] Raghunathan, T. E., Lepkowski, J. M., Hoewyk, J. V., & Solenberger, P. (2001). A multivariate technique for multiply imputing missing values using a sequence of regression models. *Survey Methodology, 27,* 85-95.

[111] Raghunathan, T. E., Reiter, J. P., & Rubin, D. B. (2003). Multiple imputation for statistical disclosure limitation. *Journal of Official Statistics-Stockholm-, 19,* 1-16.

[112] Rao, J. N. K. (1996). On variance estimation with imputed survey data. *Journal of the American Statistical Association, 91,* 499-506.

[113] Rao, J. N. K., & Shao, J. (1992). Jackknife variance estimation with survey data under hot deck imputation. *Biometrika, 79,* 811-822.

[114] Ratitch, B., & O'Kelly, M. (2011). Implementation of pattern-mixture models using standard SAS/STAT procedures. *Proceedings of Pharma-SUG 2011 (Pharmaceutical Industry SAS Users Group). Paper SP04,* Nashville.

[115] Ratitch, B., Lipkovich, I., & O'Kelly, M. (2013). *Combining analysis results from multiply imputed categorial data* No. ParmaSUG 2013 - Paper SP03)

[116] Reiter, J. (2002). Satisfying disclosure restrictions with synthetic data sets. *Journal of Official Statistics, 18,* 531-543.

[117] Reiter, J. P., Raghunathan, T. E., & Kinney, S. K. (2006). The importance of modeling the sampling design in multiple imputation for missing data. *Survey Methodology, 32,* 143.

[118] Reiter, J. P. (2007). Small-sample degrees of freedom for multi-component significance tests with multiple imputation for missing data. *Biometrika, 94,* 502-508.

[119] Robins, J. M., Rotnitzky, A., & Zhao, L. P. (1995). Analysis of semi-parametric regression models for repeated outcomes in the presence of missing data. *Journal of the American Statistical Association, 90,* 106-121.

[120] Robins, J. M., & Wang, N. (2000). Inference for imputation estimators. *Biometrika, 87,* 113-124.

[121] Royston, P. (2007). Multiple imputation of missing values: Further update of ice, with an emphasis on interval censoring. *Stata Journal, 7,* 445-464.

[122] Rubin, D. B. (1976a). Inference and missing data. *Biometrika, 63,* 581-592.

[123] Rubin, D. B. (1976b). Comparing regressions when some predictor values are missing. *Technometrics, 18*, 201-205.

[124] Rubin, D. B. (1976c). Noniterative least squares estimates, standard errors and F-tests for analyses of variance with missing data. *Journal of the Royal Statistical Society.Series B (Methodological), 38*, 270-274.

[125] Rubin, D. B. (1977). Formalizing subjective notions about the effect of nonrespondents in sample surveys. *Journal of the American Statistical Association, 72*, 538-543.

[126] Rubin, D. B. (1978). Multiple imputations in sample surveys-a phenomenological bayesian approach to nonresponse. *Proceedings of the Survey Research Methods Section of the American Statistical Association, 1*, 20-34.

[127] Rubin, D. B. (1987). *Multiple imputation for nonresponse in surveys* (99th ed.) Wiley.

[128] Rubin, D. B., & Schenker, N. (1986). Multiple imputation for interval estimation from simple random samples with ignorable nonresponse. *Journal of the American Statistical Association, 81*, 366-374.

[129] Rust, K. (1985). Variance estimation for complex estimators in sample surveys. *Journal of Official Statistics, 1*, 381-397.

[130] Schafer, J. L. (1997). *Analysis of incomplete multivariate data*. New York: Chapman & Hall.

[131] Schenker, N., Raghunathan, T. E., & Bondarenko, I. (2010). Improving on analyses of self-reported data in a large-scale health survey by using information from an examination-based survey. *Statistics in Medicine, 29*, 533-545.

[132] Schminkey, D. L., von Oertzen, T., & Bullock, L. (2016). Handling missing data with multilevel structural equation modeling and full information maximum likelihood techniques. *Research in Nursing & Health, 39*, 286-297.

[133] Schumacker, R. E., & Lomax, R. G. (1996). *A beginner's guide to structural equation modeling*. Mahwah, NJ: L. Erlbaum Associates.

[134] Seaman, S. R., Bartlett, J. W., & White, I. R. (2012). Multiple imputation of missing covariates with non-linear effects and interactions: An evaluation of statistical methods. *BMC Medical Research Methodology, 12*(1), 46.

[135] Siddiqui, O., & Ali, M. W. (1998). A comparison of the random-effects pattern mixture model with last-observation-carried-forward (LOCF) analysis in longitudinal clinical trials with dropouts. *Journal of Biopharmaceutical Statistics, 8*, 545-563.

[136] Siscovick, D. S., Raghunathan, T., King, I., Weinmann, S., Bovbjerg, V. E., Kushi, L., et al. (2000). Dietary intake of long-chain n-3 polyunsaturated fatty acids and the risk of primary cardiac arrest. *The American Journal of Clinical Nutrition, 71*(1 Suppl), 208S-12S.

[137] Siscovick, D. S., Raghunathan, T. E., King, I., Weinmann, S., Wicklund, K. G., Albright, J., et al. (1995). Dietary intake and cell membrane levels of long-chain n-3 polyunsaturated fatty acids and the risk of primary cardiac arrest. *Jama, 274*, 1363-7.

[138] Stan Development Team (2017). Stan Modeling Language Users Guide and Reference Manual, Version 2.17.0. http://mc-stan.org

[139] Stasny, E. A. (1986). Estimating gross flows using panel data with nonresponse: An example from the canadian labour force survey. *Journal of the American Statistical Association, 81*, 42-47.

[140] StataCorp. (2017). *Stata statistical software: Release 15.* College Station, TX: StataCorp, LLC.

[141] Stokes, M. E., Davis, C. S., & Koch, G. G. (2001). *Categorical data analysis using the SAS system* (2nd ed.). Cary, NC:Sas Institute Inc.: Wiley.

[142] Thompson, S. K. (2012). Sample size. *Sampling* (pp. 53-56) John Wiley & Sons, Inc.

[143] Tobin, J. (1958). Estimation of relationships for limited dependent variables. *Econometrica, 26*, 24-36.

[144] Valliant, R., Dever, J. A., & Krueter, F. (2013). *Practical tools for designing and weighting survey samples.* New York Heidelberg Dordrecht London: Springer.

[145] van Buuren, S. (2007). *Multiple Imputation of Discrete and Continuous Data by Fully Conditional Specification*

[146] van Buuren, S. (2012). *Flexible imputation of missing data.* Boca Raton, FL: Chapman and Hall/CRC.

[147] van Buuren, S., Boshuizen, H. C., & Knook, D. L. (1999). Multiple imputation of missing blood pressure covariates in survival analysis. *Statistics in Medicine, 18*, 681-694.

[148] van Buuren, S., & Oudshoorn, K. (1999). Flexible multivariate imputation by MICE. *Technical Report, Leiden: TNO Preventie En Gezondheid, TNO/VGZ/PG 99.054.*

[149] Verbeke, G., & Molenberghs, G. (2000). *Linear mixed models for longitudinal data* Sprinter Verlag.

[150] Von Hippel, P. T. (2007). Regression with missing YS: An improved strategy for analyzing mulitply imputed data. *Sociological Methodology, 37,* 83-117.

[151] Von Hippel, P. T. (2009). How to impute interactions, squares, and other transformed variables. *Sociological Methodology, 39,* 265-291.

[152] Weisberg, S. (2013). *Applied linear regression* (4th ed.) Wiley.

[153] White, I. R., Royston, P., & Wood, A. M. (2011). Multiple imputation using chained equations: Issues and guidance for practice. *Statistics in Medicine, 30,* 377-399.

[154] White, I. R., & Royston, P. (2009). Imputing missing covariate values for the cox model. *Statistics in Medicine, 28,* 1982-1998.

[155] Xie, X., & Meng, X. (2017). Rejoinder please visit the wild arboretum of multi-phase inference. *Statistica Sinica, 27,* 1584-1594.

[156] Yang, S., & Kim, J. K. (2016). Fractional imputation in survey sampling: A comparative review. *Statist.Sci.,* 31(3), 415-432.

[157] Zhang, W., & Yung, Y. (2011). A tutorial on structural equation modeling with incomplete observations: Multiple imputations and FIML methods using SAS. *International Meeting of Pyschometric Society,* Tai Po, Hong Kong.

[158] Zhou, H., Elliott, M. R., & Raghunathan, T. E. (2016a). A two-step semiparametric method to accommodate sampling weights in multiple imputation. *Biometrics, 72,* 242-252.

[159] Zhou, H. (2014). Accounting for complex sample designs in multiple imputation using the finite population bayesian bootstrap. (Doctoral Dissertation, University of Michigan, 2014). (2027.42/108807)

[160] Zhou, H., Elliott, M. R., & Raghunathan, T. E. (2016b). Multiple imputation in two-stage cluster samples using the weighted finite population bayesian bootstrap. *Journal of Survey Statistics and Methodology, 4,* 139-170.

[161] Zhou, H., Elliott, M. R., & Raghunathan, T. E. (2016c). Synthetic multiple-imputation procedure for multistage complex samples. *Journal of Official Statistics, 32,* Mar 2016.

[162] Zhou, X., & Reiter, J. P. (2010). A note on Bayesian inference after multiple imputation. *The American Statistician, 64,* 159-163.

[163] Zhu, J. (2016) Assessment and Improvement of a Sequential Regression Multivariate Imputation Algorithm. Ph.D. Disertation, University of Michigan, Ann Arbor, Michigan, USA.

[164] Zhu, J. & Raghunathan, T. E. (2015).Convergence Properties of a Sequential Regression Multiple Imputation Algorithm. *Journal of American Statistical Association*, 110, 1112-1124.

Index

Printed and bound by CPI Group (UK) Ltd, Croydon, CR0 4YY

24/10/2024

01778306-0005